T0331220

Guidelines for Process Safety in Bioprocess Manufacturing Facilities

Guidelines for Process Safety in Bioprocess Manufacturing Facilities

Center for Chemical Process Safety
New York, New York

An AIChE Technology Alliance

Center for Chemical Process Safety

A JOHN WILEY & SONS, INC., PUBLICATION

A Joint Publication of the Center for Chemical Process Safety of the American Institute of Chemical Engineers and John Wiley & Sons, Inc.

Published by John Wiley & Sons, Inc., Hoboken, New Jersey.
Published simultaneously in Canada.

For general information on our other products and services or for technical support, please contact our Customer Care Department within the United States at (800) 762-2974, outside the United States at (317) 572-3993 or fax (317) 572-4002.

Wiley also publishes its books in a variety of electronic formats. Some content that appears in print may not be available in electronic format. For information about Wiley products, visit our web site at www.wiley.com.

Library of Congress Cataloging-in-Publication Data:

Guidelines for process safety in bioprocess manufacturing facilities / Center for Chemical Process Safety of the American Institute of Chemical Engineers.
   p. cm.
  Includes index.
  ISBN 978-0-470-25149-2 (hardback)
  1. Biochemical engineering—Safety measures. I. American Institute of Chemical Engineers. Center for Chemical Process Safety.
  TP248.3.G85 2011
  660.6028'9—dc22   2010036824

Printed in the United States of America

oBook ISBN: 978-0470-94914-6
ePDF ISBN: 978-0470-94913-9

10 9 8 7 6 5 4 3 2 1

CONTENTS

List of Tables ....*xi*

List of Figures...*xiii*

Items on the Web Accompanying This Book...*xv*

Acknowledgements...*xvii*

Preface..*xix*

1   INTRODUCTION   1

1.1	Bioprocess Engineering Information Transfer and Management Practices	3
1.2	The Need for Bioprocess Safety Management Systems	7
1.2.2	Bioprocessing Incidents and Releases	8
1.3	Our Target Audience	14
1.4	How to use this Guideline	15

2   AN OVERVIEW OF THE BIOPROCESSING INDUSTRY   17

2.1	Bioprocessing's History	17
2.1.1	Bioprocessing's Historical Advancement	18
2.1.1.1	Microbiological Advancements	18
2.1.1.2	Food Science and Food Process Technology Advancements	19
2.1.1.3	Genetic Advancements	19
2.1.1.4	Future Bioprocessing Developments	20
2.2	Industrial Applications	20
2.2.1	Processes	21
2.2.2	Products	21
2.3	The Bioprocess Lifecycle	22

2.3.1   Discovery   23
2.3.2   Development Phase: Laboratory and Pilot Plant   23
2.3.3   Scale-up Phase   24
2.3.4   Upstream Operations and Downstream Operations   26
2.3.4.1 Inoculation / Seed and Production Biosafety
         Containment and Production Risk   27
2.3.4.2 Fermentation / Cell Culture   31
2.3.4.3 Scale of Manufacturing   36
2.3.5   General Biosafety Recommendations for Large Scale
         Work   38
2.3.5.1 Facility Design   39
2.3.5.2 Equipment Design   39
2.3.5.3 Cleaning, Inactivation, and Sterilization   41
2.3.5.4 Maintenance   42
2.3.5.5 Air and Gas Emissions   42
2.3.5.6 Waste Handling   42
2.3.5.7 Accidental Release   43
2.3.6   Product Safety Information   43
2.3.6.1 Product Handling   44
2.3.6.2 Material Disposal   44
2.3.6.3 Disposable Process Technology   44
2.3.7   Outsourced Manufacturing Concerns   45

3   BIOPROCESSING SAFETY MANAGEMENT PRACTICES   47

3.1   Sample Approach   48
3.1.2   Develop and Document a System to Manage
         Bioprocess Safety Hazards   50
3.1.3   Appoint a Biological Safety Officer   50
3.1.4   Collect Bioprocess Hazard Information   51
3.1.5   Identify Bioprocess Safety Hazards   51
3.1.5.1 Point of Decision   51
3.1.6   Assess Bioprocess Safety Risks and Assign
         Bioprocess Safety Hazard Level   52
3.1.7   Identify Bioprocess Controls and Risk Management
         Options   52
3.1.8   Document Bioprocess Safety Hazard Risks and
         Management Decisions   53
3.1.9   Communicate and Train on Bioprocess Safety Hazards   53

	3.1.10	Investigate & Learn from Bioprocess Incidents	53
	3.1.11	Review, Audit, Manage Change, and Improve Hazard Management Practices and Program	54
	3.2	Existing Management Systems	54
	3.2.1	Product Stewardship for Bioproducts	61
	3.3	Establishing a Bioprocess Safety Management System	62
	3.3.1	Select a Management System Model Based Upon Your Needs	63
	3.3.2	Identifying the Elements that Apply to Your Operations	64
	3.3.3	Establish a Review and Approval Cycle for the Documents	65
	3.3.4	Rolling Out the Management System to the Users	66
	3.4	Biosafety Training for the Workforce	67
	3.5	Investigating Incidents	69
	3.5.1	A Generic Procedure for Initial Biohazard Incident Response	71
	3.6	Managing Change	75
	3.7	Reviewing and Auditing for Continuous Improvement	76
	3.8	Applying Behavior-Based Safety to Bioprocesses	76

4 IDENTIFYING BIOPROCESS HAZARDS 79

	4.1	Key Considerations for Assessing Risk to Manage Bioprocess Safety	79
	4.1.1	Testing for Bioactivity	79
	4.1.2	Non-biological Hazards	80
	4.2	Bioprocess Risk Assessment	80
	4.2.1	Three Types of Assessment	80
	4.2.2	Agent Considerations	80
	4.2.3	Process Considerations	81
	4.2.4	Environmental Considerations	82
	4.2.5	Microorganisms	83
	4.3	Recombinant Organisms	85
	4.4	Cell Culture	86

5 BIOPROCESS DESIGN CONSIDERATIONS AND UNIT OPERATIONS 89

	5.1	Physical Plant Design	89
	5.1.1	Architectural Aspects	90

5.1.1.1	Finishes and Materials	90
5.1.1.2	Layout Strategies	91
5.1.1.3	People and Material Flow	94
5.1.1.4	Non-biological Hazards	94
5.1.1.5	Seismic and Building Loads	96
5.1.1.6	Hardened Construction	97
5.1.1.7	Equipment Mezzanines and Subfloors	97
5.1.1.8	Heating, Ventilation, and Air Conditioning Aspects	98
(a)	Supply and Exhaust Systems	98
(b)	Special Exhaust Stream Mitigation	100
(c)	HVAC Issues from a Biosafety Perspective	101
(d)	Microenvironments	103
(e)	Cascading Pressure Differentials	105
(f)	Containment versus Clean Room Environments	107
5.1.1.9	Waste and Waste Treatment	109
5.1.1.10	Process Support Systems: High Purity Water	112
5.1.1.11	Process Support Systems: Hand Washing Sinks and Personnel showers	112
5.1.2	Plant Siting Issues	113
5.1.2.1	Zoning & Permitting	113
5.1.2.2	Regional Environmental Agencies and Environmental Impact Reports	113
5.1.2.3	Building and Site Security	114
5.2	Bioprocess Unit Operations	116
5.2.1	General Equipment Design Considerations	117
5.2.2	Closed-System Design	118
5.2.2.2	Impact on Operations	123
5.2.3	Upstream Equipment and Facility Design	124
5.2.3.1	Additional Upstream Design Considerations	124
5.2.3.2	Equipment and Facility Integration	127
5.2.3.3	Production Segregation and Flows	127
5.2.3.4	Segregation from a Biosafety Perspective	129
5.2.3.5	Cleaning the Equipment	130
5.2.4.1	Harvest and Recovery	134
5.2.4.2	Centrifugation	134
5.2.4.3	Filtration	135
5.2.4.4	Chromatography	137
5.2.5	Facility Support Issues	139
5.2.6	Biosafety for Personnel: SOP, Protocols, and PPE	140

6 THE EFFECTS OF EMERGING TECHNOLOGY ON BIOPROCESSING RISK MANAGEMENT 143

6.1	Researching and Staying Informed	143
6.1.1	Biopharmaceutical	144
6.1.1.1	Drug Discovery and Development	144
6.1.1.2	Gene-based Pharmaceuticals	144
6.1.1.3	Drug Delivery Research	146
6.1.2	Renewable-resources	147
6.1.3	Environmental	148
6.1.3.1	Bioprocessing and Waste Management	148
6.2	Communicating the Impacts of New Technology	149
6.2.1	Industry (Communication at Your Site)	150

APPENDIX A – REFERENCES & SELECTED REGULATIONS   153

APPENDIX B – LARGE SCALE BIOSAFETY GUIDELINES   161

APPENDIX C – A GENERIC LABORATORY/LARGE SCALE BIOSAFETY CHECKLIST   177

APPENDIX D – BIOLOGICAL ASSESSMENT QUESTIONNAIRE & BIOPROCESS SAFETY CHECKLIST   179

APPENDIX E – BIOPROCESS FACILITY AUDIT CHECKLIST   189

APPENDIX F – DIRECTIVE 2000/54/EC OF THE EUROPEAN PARLIAMENT AND OF THE COUNCIL   199

APPENDIX G – COMPARISON OF GOOD LARGE SCALE PRACTICE (GLSP) AND BIOSAFETY LEVEL (BL) - LARGE SCALE (LS) PRACTICE   203

GLOSSARY   209

ACRONYMS AND ABBREVIATIONS   217

INDEX   221

LIST OF TABLES

TABLE 1-1   Typical Hazards at Bioprocess Manufacturing Sites   4

TABLE 1-2   Preliminary Design Anticipated Hazards Analysis   6

TABLE 1-3   Incidents Related to Products of Biologic Origin   10

TABLE 1-4   Selected International Biosafety Levels and Acronyms   13

TABLE 2-1   Comparison of Process Parameters and Characteristics of
Recombinant Bacterial and Mammalian Cell Lines for
Production of Recombinant DNA Protein Products   33

TABLE 2-2   Components of Typical Growth Media   36

TABLE 2-3   Examples of Very Large Scale Bioprocessing Products   37

TABLE 3-1   Comparison of a CEN-Based Biorisk Management System
to OSHA PSM, EPA RMP, and CCPS Risk-Based Process
Safety (RBPS)   60

TABLE 3-2   A Five-phase Review and Approval Cycle   66

TABLE 4-1   WHO Risk Group Classifications   83

TABLE 5-1   Clean Room Classifications   99

LIST OF FIGURES

Figure 1-1 The International Biohazard Symbol 8

Figure 2-1 Typical Process Operations for a Bioprocess 24

Figure 2-2 Typical Steps to Develop rDNA *E. coli* 25

Figure 2-3 Typical Steps to Manufacture Human Insulin 25

Figure 2-4 Basic Biotech Manufacturing Unit Operations 26

Figure 2-5 The Cell Culture Scale-Up Process (Staging) 28

Figure 2-6 A Common Spinner Flask 29

Figure 2-7 Seed Bioreactors 30

Figure 2-8 The Cell Growth Lifecycle 32

Figure 2-9 Stirred-Tank Bioreactor 40

Figure 2-10 Large Scale Wave Bioreactor (Courtesy of GE Healthcare) 41

Figure 3-1 Bioprocess Hazard Management Implementation Flowchart 49

Figure 5-1 Nested Zones 93

Figure 5-2 People and Material Flow Diagram 94

Figure 5-3 Vaporized Hydrogen Peroxide (VHP) decontamination unit 96

Figure 5-4 A Wet Scrubber 103

Figure 5-5 Image of a Microenvironment 104

Figure 5-6 A Typical Glove Box 105

Figure 5-7 HEPA Filter Installation 106

Figure 5-8 Air Lock Types Pressure Differential Diagram 107

Figure 5-9 Full Body Positive Pressure Suit 108

Figure 5-10 Effluent Decontamination System 111

Figure 5-11 Retinal Scanning Device 115

Figure 5-12 Closed Disposable Sampling System 123

Figure 5-13 Traditional Clean Room Approach 125

Figure 5-14 Gray Space Approach 126

Figure 5-15 Minimizing Classified Space 127

Figure 5-16 Flow and Segregation Relationships 129

Figure 5-17 Clean-in-Place Hard Spots 131

Figure 5-18 Disk-type Centrifuge   135
Figure 5-19 Ultrafiltration Skid (Courtesy of Sartorius)   136
Figure 5-20 Chromatography Column Schematic (Courtesy of Optek)   138

FILES ON THE WEB ACCOMPANYING THIS BOOK

Biological Assessment Questionnaire

Bioprocess Facility Audit Checklist

You can access these files by going to the site:

www.aiche.org/ccps/publications/bioprocess.aspx

To access the files, download the zipped folder and extract all of the files. You will be asked for a password, enter the password:

CCPSBio2010

If you have difficulty accessing the files, contact CCPS at ccps@aiche.org or +1.646.495.1371

ACKNOWLEDGEMENTS

The American Institute of Chemical Engineers (AIChE) wishes to thank the Center for Chemical Process Safety (CCPS) and those involved in its operation, including its many sponsors whose funding made this project possible, and the members of the Technical Steering Committee, who conceived of and supported this guideline project. The members of the bioprocess safety management subcommittee who worked with AntiEntropics, Inc. to produce this text deserve special recognition for their dedicated efforts, technical contributions, and overall enthusiasm for creating a useful addition to the process safety guideline series. CCPS also wishes to thank the subcommittee members' respective companies for supporting their involvement in this project as well as the American Biological Safety Association (ABSA) for creating a source for sharing valuable bioprocessing safety information and research.

The chairman of the bioprocess safety management subcommittee was Will Fleming of Bristol-Myers Squibb. The CCPS staff liaison was Dan Sliva. The members of the CCPS guideline subcommittee were:

- Buddy Bowman, Syngenta
- Rick Braun, IPS
- Mary Cipriano, Abbott Laboratories
- Aaron Duff, Bristol-Myers Squibb
- Bruce Greer, Scientific Protein Laboratories LLC
- Jose Hanquier, Eli Lilly and Company
- Jerry Jones, Genentech
- Beth Junker, Merck & Company, Inc.
- Chantel Laing, Schering-Plough Corporation
- Denise Lackey, Amgen, Inc.
- Richard Medwid, Eli Lilly and Company
- Barbara Owen, Bristol-Myers Squibb
- Alan Powell, Merck & Company, Inc.
- Robert Stankovich, Eli Lilly and Company

AntiEntropics, Inc. of New Market, Maryland, was the contractor for this project. Robert J. Walter was the principal co-author and project manager. Sandra A. Baker was co-author and editor. Joseph Kallhoff was a contributing author. In addition AntiEntropics would like to recognize the following contributors:

- Bob Stankovich of Eli Lilly and Company for his co-authorship of Chapter 3
- Mary Cipriano of Abbott Laboratories for her authorship of Chapter 4
- David McGlashan of Caris DX and Jeffery Odum of NC BioSource for contributing Chapter 5.

CCPS also gratefully acknowledges the comments submitted by the following peer reviewers:

- William Gaylord, Allergan
- Robert Kiss, Genentech
- David R. Maraldo, Ph.D., Merck & Company, Inc.
- Chris Meyer, Eli Lilly and Company
- Dan Noberini, Bristol-Myers Squibb
- Venkata Ramana, Reliance Life Sciences Pvt. Ltd,
- Richard Rebar, GlaxoSmithKline
- Stephen Sykes, United States Food and Drug Administration
- Dan Wozniak, Abbott Laboratories
- Timothy E. Woenker, Chematics, Inc.

Their insights, comments, and suggestions helped ensure a balanced perspective for this guideline.

PREFACE

The American Institute of Chemical Engineers (AIChE) has been closely involved with process safety and loss control issues in the chemical and allied industries for more than four decades. Through its strong ties with process designers, constructors, operators, safety professionals, and members of academia, AIChE has enhanced communications and fostered continuous improvement of the industry's high safety standards. AIChE publications and symposia have become information resources for those devoted to process safety and environmental protection.

AIChE created the Center for Chemical Process Safety (CCPS) in 1985 after the chemical disasters in Mexico City, Mexico, and Bhopal, India. The CCPS is chartered to develop and disseminate technical information for use in the prevention of major chemical accidents. The center is supported by more than 80 chemical process industries (CPI) sponsors who provide the necessary funding and professional guidance to its technical committees. The major product of CCPS activities has been a series of guidelines to assist those implementing various elements of a process safety and risk management system. This book is part of that series.

AIChE recognized a significant increase in members' bioprocess related needs in the early 1990s. Some of these members' processes benefit from traditional process safety techniques, others present different challenges for managing the biological nature of their hazards and associated risks, and still others combine both categories of hazards. Bioprocess safety management meshes the lessons learned from over 24 years of chemical process safety management with the unique approaches demanded by the widening variety of bioprocessing safety challenges. The CCPS Technical Steering Committee initiated the creation of these guidelines to assist bioprocessing facilities in meeting these challenges. This book contains approaches for designing, developing, implementing, and continually improving a bioprocess safety management system. The website accompanying this book contains resource materials and support information.

1
INTRODUCTION

The following definition sets the scope of our discussion of process safety management in the bioprocess manufacturing industry:

> *Bioprocess—A process that makes use of microorganisms, cells in culture, or enzymes to manufacture products or complete a chemical transformation.*

Humans have been using such processes for baking bread, making cheese and fermenting alcoholic beverages since prehistoric times. Advances in commercializing recombinant DNA technology allow the production of an enormous variety of protein-based therapeutic drugs that is having a profound impact on the quality of life for severely ill patients. Bioprocessing is also essential to several emerging industries and technologies, including the production of biofuels from renewable biomass feedstocks such as ethanol biodiesel, and for the production of polymeric materials. Therapeutic stem cells, gene therapy vectors, and new vaccines are all the results of bioprocessing technology.

Effective process safety management in bioprocess manufacturing is essential to the growth of an already booming segment of global manufacturing. In the past few decades, leaps in basic science, new bioprocessing discoveries, technological methods, and equipment design have created a vibrant and creative business segment. Bioprocessing is a business segment that, like any other, has traditional fiscal risks but then adds unique chemical and biological hazard-based risks related to

- the raw materials involved,
- the products made,
- the processes used,
- the waste streams involved, and
- unique end user considerations.

1

Effective process safety management is viewed worldwide by leaders in the chemical process industries, government regulatory agencies, and non-governmental public advocacy groups concerned with public safety and environmental protection as a business philosophy that supports safe, efficient, and reliable operation of manufacturing facilities. An increased emphasis on process safety management across many segments of the process industries during the last several decades is widely credited for reducing the risks of catastrophic accidents in facilities worldwide.

While process safety management has traditionally been focused upon large facilities in the petroleum, natural gas and chemicals and polymers production sectors, other facilities in the process industries have also widely used and benefited from the basic concepts of process safety. Examples include facilities that may not be required by regulations to adopt formal process safety management systems such as biopharmaceutical industry facilities for production of biological drug substances and vaccines.

This book addresses process safety management practices for manufacturing facilities that use bioprocesses. For the purposes of this guideline, the reader should expand the traditional scope of the definition of hazardous materials to include chemicals, biological agents, and intermediates and derivatives generated during manufacture.

Owners of bioprocessing facilities must manage a variety of process safety related hazards, not just biohazards. These include a variety of chemical hazards and physical hazards (for example, stored energy in pressure vessels located in utility supply and process areas, asphyxiant gases, hot acidic and caustic cleaning solutions). While biohazards may in some instances be very significant and perhaps of primary importance from a risk perspective, in many cases chemical and physical hazards will present the more significant risk exposures to workers, neighbors, the environment, and property.

This book is a survey of the present guidance and experience from industry, professional organizations, and governmental research to encourage and support the use of systematic and self-directed design for success in

- safety (including bioprocess safety, personnel safety, and chemical process safety),
- environmental responsibility,
- quality, and
- the business case for your organization to embrace a rigorous management system to support all of the above as they apply to your organization.

1.1 BIOPROCESS ENGINEERING INFORMATION TRANSFER AND MANAGEMENT PRACTICES

A smooth interface between bioprocess scientists, bioprocessing engineers, biosafety specialists, and technical and support professionals demands a management system to address the transfer and consistent application of technology—both process technology and safety technology—from the laboratory to the production floor. Success in achieving this goal depends upon the combination of well understood bioprocessing guidelines and regulatory compliance methods with proven safety management, bioprocessing management, and business management best practices. These include, but are not limited to:

- Occupational Safety and Health Administration (OSHA) process safety management techniques
- Food and Drug Administration (FDA) Good Manufacturing Practice guidelines
- National Institutes of Health (NIH) guidance for facility design and specific bioactive and biological material use
- Center for Disease Control (CDC) and World Health Organization (WHO) laboratory safety management guidance
- Organisation for Economic Co-operation and Development (OECD) Directive on the protection of workers from risks related to exposure to biological agents at work
- United States Department of Agriculture (USDA) facility design standards
- Environmental Protection Agency (EPA) risk management program techniques
- International Conference on Harmonization of Technical Requirements for Registration of Pharmaceuticals for Human Use (ICH).
- Professional society standards and practices, such as the International Society for Pharmaceutical Engineering (ISPE), American Society for Microbiology (ASM), American Society for Mechanical Engineers (ASME), American Biological Safety Association (ABSA)
- Integrated operational excellence business management system techniques

This book presents the concept of process safety for bioprocesses as a branch within a total business management system. Operational excellence is supported when the various arms of the business all follow similarly detailed management systems.

TABLE 1-1 Typical Hazards at Bioprocess Manufacturing Sites

Asphyxiant Gases	Forklifts & Aerial Lifts
Biohazards	Mechanical: Rotating Shafts, Pinch Points, Entrapment, Shop Tools, & Hand Tools
Chemicals Used for Equipment and Work Surface Cleaning and Maintenance	Natural Perils: Wind, Floods, Extreme Temperatures, Earthquakes
Chemicals Used in Product Synthesis, Recovery, Purification, and Waste Treatment	Noise
Confined Spaces	Pharmaceutical Active Product Exposures
Cranes & Hoists	Pressure Vessels: Fired and Unfired
Electrical	Slips & Trips
Ergonomic	Thermal: Steam, Hot Water, Refrigerant, & Cryogenic Fluids
Explosions (due to powders or flammable solvents)	Wastes for Offsite Disposal: Solid & Liquid
Falls from Height	Wastes: Emissions to Air
Fires	Wastewater Discharges

Manufacturing facilities using bioprocesses represent a variety of traditional workplace physical and chemical hazards but add potential biological hazards. The severity or magnitude of these various hazards depends upon

- the type of products being manufactured,
- the types of chemical, physical, and bioprocesses used to manufacture the product and the types of raw materials and processing aids or solvents,
- the specific type of biological organism or cellular products used,
- co-products or by-products produced and the physical and chemical processes required for product recovery and purification,
- yields and waste streams generated per unit of product produced, and
- the scale of the operation and schedule for production (for example, short campaigns for products that change often versus continual production during most of the year).

As illustrated by Table 1-1, *Typical Hazards at Bioprocess Manufacturing Sites*, many of the physical, mechanical, and chemical hazards faced in a bioprocessing facility are comparable to those addressed in chemical synthesis and processing plants. Of course the major difference between a bioprocess plant and a chemical plant is the potential for biohazards. In some cases the biohazards represent extremely low risks (for example, most recombinant mammalian cell lines used for large scale production of antibody and protein drugs). However, in cases where infectious organisms are used, or where the culture may be susceptible to adventitious contamination (for example, contamination of human cell lines with a virus), the hazard may be much more significant and the risks to workers or the public from an accidental release considerably higher. The following shows an excerpt from a process hazard analysis conducted during a preliminary design review of hazards for a mammalian cell culture process. The analysis was conducted during an early phase of the design, and each area of the process was analyzed for both biological and traditional safety issues. Consider this approach for your facility.

TABLE 1-2 Preliminary Design Anticipated Hazards Analysis

Anticipated Hazards	Unit Operation or Process Step with Opportunity for Exposure
Exposure / release of recombinant CHO cell line	Spills Sampling Filter changes Vent stacks Non-routine operations (for example, access biokill system for maintenance) Lab sampling
Exposure/ release of biological toxins	On-site wastewater treatment facility Process filter systems Centrifuge/cell lysing operations
Exposure to acid / caustic chemicals	Clean in place (CIP) bulk chemical distribution Wastewater treatment plant (WWTP) neutralization Ammonia for process pH control
Exposure to powders (media prep and buffer prep)	Powders not fully characterized for risks by manufacturers Inorganic salts (low risk)
Exposure to asphyxiants gas	Carbon dioxide for cell culture Nitrogen for cell banks
Exposure to cytotoxics	Methotrexate for cell culture
Process Powders – combustible dust considerations / exclusivity potential	Buffer and Media preparation Dispensing
Pressure	Steam use in processing Transfer panels requiring manual intervention Opening tanks or closed process equipment for inspection and swabbing Filter changes
High temperatures	Steam use in sterilization Autoclaves Water for injection (WFI) point of use drops
Freezing temperatures	Cryogenic cylinder use Cell bank storage Walk-in freezers for final products
Noise	Centrifuge Chillers Air Compressors High velocity gasses (steam/air) sparged into vessels or in exhausts Mechanical areas

Anticipated Hazards	Unit Operation or Process Step with Opportunity for Exposure
Lab chemicals	Analyses
Process flammable liquids	ethanol for purification column anti-bactericide Column resins delivered in alcohol preservative solutions
Slips & Trips	High cleanability surfaces create risk Wet floors, condensate leaks, mopping
Microbial control chemicals	Hydrogen peroxide, Superchlor, and other agents
Fire	Minimal process fire risks
Confined Space Entry	Validation Inspection Sampling/Swabbing Maintenance
LOTO	Validation Inspection Sampling/swabbing Maintenance
Line Breaking	Validation Inspection Cleaning
Ergonomics	Media and buffer prep solutions Portable process equipment Product containers

1.2 THE NEED FOR BIOPROCESS SAFETY MANAGEMENT SYSTEMS

The hazard and risk exposure presented in bioprocessing industry operations establishes a solid case for developing a culture of bioprocessing safety management systems. Clear instructions for the physical operations, maintenance tasks, and administrative workflow processes are essential to reproducible quality assurance and quality control metrics.

A custom management system for a facility that incorporates process safety management (PSM) system practices appropriate to the biohazards and risks as well as other specific hazards and risks they represent to workers at the site, the environment, and the surrounding community provides the following benefits to the facility:

- For companies with multiple sites, it can help ensure that a minimum corporate structure is in place which can be customized for differing sites.
- For a single facility, it can establish a model for growth and a way to capture organizational requirements to achieve that growth.

- An integrated bioprocess safety management system allows controlled modifications to reflect technology changes as well as day-to-day processing step changes.
- It provides a specific approach for the site's biosafety management system using a risk-based approach based upon the hazards of the process.

Figure 1-1   The International Biohazard Symbol

Figure 1-1, *The International Biohazard Symbol*, was developed by the Dow Chemical Company in 1966 while developing containment systems for the Cancer Institute at the National Institutes of Health. Mr. Charles Baldwin, an environmental health engineer, participated in its development. He reported that his team saw a need for this kind of a symbol as there was no standardization at the time. They proceeded to develop some symbols with the help of the Dow marketing department.

To quote Mr. Baldwin from an interview with the *New York Times*, "We wanted something that was memorable but meaningless, so we could educate people as to what it means." The resulting symbol is now found in almost every location where biohazardous material is handled or stored.

1.2.2   Bioprocessing Incidents and Releases

Over the past 50 years, pharmacological products of biologic origin have resulted in incidents that emphasize the need for careful quality and biosafety assessment of the production processes. Physical and chemical hazards in facilities that use bioprocesses are often comparable to those found in other sectors of industry. Facilities using bioprocesses for environmental control purposes may use flammable solvent as reagents or methane-rich fuel gas may be produced as a plant by-product from waste degradation under anaerobic conditions. In the case of the production of fuels from biomass feedstock using bioprocesses, there are numerous areas of the facilities where significant fire and explosion hazards are present.

The biofuels industry includes many bioprocess facilities that produce ethanol using a yeast fermentation process that converts starch-derived sugars to ethanol. The hazards associated with such facilities that utilize grain crops as the starch source for sugars have been described in the literature. These include:

- Potential releases of hazardous chemicals such as anhydrous ammonia (used as a nitrogen source in fermentation)
- Potential fires or explosions associated with
    o Grain dust in grain storage, handling and processing, and
    o Processing and storing flammable liquids (ethanol solutions in flammable concentration range).

Table 1-3, *Incidents Related to Products of Biologic Origin*, summarizes some major bioprocess-related incidents. These incidents affected four categories of victims: laboratory workers, production workers, the general populace, and patients treated with adulterated pharmaceuticals. These examples define the need for managing process safety in bioprocessing manufacturing facilities.

TABLE 1-3 Incidents Related to Products of Biologic Origin

Incident	Comments
Incomplete virus inactivation	**Polio vaccine**: In 1955 the "Cutter incident" occurred in which 200,000 people were inadvertently injected with live virulent polio virus; 70,000 became ill, 200 were permanently paralyzed, and 10 died. **Rabies vaccine**: In 1960, the "Fortaleza incident" occurred resulting in an outbreak of post-vaccinal rabies (rage de laboratoire) in Fortaleza, Brazil.
Endogenous viral contamination	The **yellow fever immunization** during World War II used vaccines produced from chick embryos that almost certainly contained avian leucosis-sarcoma viruses. Some of the **polio vaccine** administered from 1955–1963 was contaminated with a virus called simian virus 40 (SV40). SV40 was found in the primary Rhesus monkey kidney cell line used in the virus manufacturing bioprocess.
Adventitious viral contamination	In the 1940s **Hepatitis B (HBV) contaminated serum** was used as stabilizer in yellow fever vaccines. Many patients developed HBV. Recently, Parvovirus B19 contaminated **albumin immunoaffinity-purified factor VIII concentrate** (an antihemophilic). Recombinant FVIII Parvovirus B19 (B19) is known to cause a variety of human diseases in susceptible individuals.
Infected source material or inadequate virus clearance	Recently, **plasma derived products** have been found to be sources of HIV, hepatitis virus, and parvovirus B19 transmission.
Accidental worker exposure	In the mid 1990s accidental exposure to **livestock brucellosis vaccine RB 51** occurred. Of 26 cases 21 were by needle stick injury, four by conjunctival spray exposure, and one by spray exposure of an open wound.
Accidental worker exposure	General exposure from **wastewater treatment plants** at bioprocessing facilities exposures have been found to result from the inherent open operations. Aeration tanks, sampling systems, and maintenance operations can all result in contact with wastes.
Public environmental exposure	In 1979, an unusual **anthrax (*Bacillus anthracis*)** epidemic occurred in Sverdlovsk, Union of Soviet Socialist Republics. It was concluded that the escape of an aerosol of anthrax pathogen at a nearby military facility caused the outbreak.
Three workers injured in explosion	In 2008, a **bioprocess ethanol end product** exploded due to process safety issues.

In addition to the toxic, flammable, or explosive process safety risks that may be present at a bioprocessing facility, there are additional factors to consider when planning a comprehensive safety management system for processes involving biohazardous or potentially biohazardous materials. These are as follows:

- **Biological Agent:**
  - o Pathogenicity
  - o Infectious dose
  - o Virulence (primary or secondary communicability)
  - o Host factors (immunocompetence, pregnancy, underlying medical conditions, extreme age, or immunity)
  - o Sensitization reactions (allergens, toxins, or biologically active compounds)
  - o Incidence of laboratory acquired infections (LAI)
  - o Availability of vaccine and / or prophylactic treatment
  - o Environmental impact (Agent stability—sensitivity to chemical and physical inactivation—survivability and dissemination in the environment)
- **Routes / Modes of Transmission in the Workplace**
  - o Respiratory: Inhaling of contaminated particles
  - o Mucous membrane: Splashing, spraying or droplets in the eyes or mouth
  - o Parenteral: Penetrations through the skin; for example, cuts, needle sticks, or abrasions
  - o Non-intact skin: Contact with skin affected with dermatitis, chafing, hangnails, abrasions, acne, or other conditions that can alter the barrier properties of the skin
  - o Ingestion: Swallowing contaminated material
  - o Adsorption: Adhesion to a surface

- **Environmental factors**
  - o Climate
  - o Geography
  - o Proximity to people
- **Procedural and facility factors**
  - o Ventilation and laboratory design: Directional air, pressure gradients, airbreaks, separation of laboratories from offices, interlocking autoclave and airlock doors
  - o Laboratory procedures: Use of inherently safer engineered sharps, containment of aerosols, and other means
  - o Containment equipment: Class II and III biological safety cabinets, sealed centrifuges cups and rotors, gasket seals and unbreakable tubes
  - o PPE: Gloves, safety glasses, lab coats, face masks, respirators, or gowns
  - o Training: Standard microbiological practices, aseptic practices, decontamination, spill cleanup, and handling of accidents
  - o Facility sanitation: Decontamination, housekeeping, routine cleaning and disinfection, pest and rodent control program
  - o Medical Surveillance Monitoring: As dictated by the risks present in the bioprocessing facility

Whether traditional process safety hazards are present or not, protecting employees, the public, and the consumer equally by assessing and addressing these factors must be the prime goal of the management system. Table 1-4, *Selected International Biosafety Levels and Acronyms,* shows these acronyms and their international counterparts.

TABLE 1-4 Selected International Biosafety Levels and Acronyms

Laboratory Setting

Hazard Potential	US NIH	US CDC	CA Health Canada	UK GMO	Germany BG Chemie	UK HSE	WHO	Japan	Swiss
Very Low	BL1	BSL-1	CL-1	ACDP-1	L1	CL1	BSL1	1	1
Low	BL2	BSL-2	CL-2	ACDP-2	L2	CL2	BSL2	2	2
Moderate	BL3	BSL-3	CL-3	ACDP-3	L3	CL3	BSL3	3	3
High	BL4	BSL-4	CL-4	ACDP-4	L4	CL4	BSL4	4	4

Production Setting

Hazard Potential	US NIH	US ASM	CA Health Canada	UK GMO	Germany BG Chemie	Germany GenTSV	OECD	Japan	Swiss
Very Low	GLSP	GLSP	CL1-LS	GILSP	P1	LP0	GILSP	GILSP	NA
Low	BL1-LS	BSL1-LS	CL2-LS	GMO-B1	P2	LP1	C1	2	1
Moderate	BL2-LS	BSL2-LS	CL3-LS	GMO-B2	P3	LP2	C2	3	2
High	BL3-LS	BSL3-LS	NA	NA	P4	LP3	C3	4	3

1.3   OUR TARGET AUDIENCE

This book is intended for anyone interested in developing a new safety management system for their bioprocessing facility or upgrading or integrating an existing system. New biosafety officers and bioprocess engineers needing a resource on biosafety will find this guideline helpful. It can also serve as a primer for anyone new to the large scale bioprocessing design and operations industry. These readers might benefit from an overview of the safety management issues. It offers examples of bioprocess safety management techniques used by industry leaders and summarizes recognized reference material used throughout the world. If your facility has a mature management system in place, it can be used to check your practices and help reinforce or improve upon the current methodology. This book does not focus on live animal or agricultural biosafety, although some of the techniques and management system approaches may apply.

The following list identifies typical positions within an organization that may benefit from this guideline:

- Bioprocess safety professionals at the plant and corporate levels
- Process safety and risk management program managers and coordinators at manufacturing facilities where these regulations apply
- Corporate bioprocess safety management staff
- Project managers and project team members whose projects initiate the need to implement a bioprocess safety management system or modify an existing system to reflect changes in processing technology or expansions
- Engineers or other staff members performing management of change activities
- Engineers modifying facilities or building new facilities
- Operations, maintenance, and other manufacturing personnel who may be critical in implementing parts of a facility's bioprocess safety management system
- Plant managers and other members of manufacturing plant leadership teams responsible for the overall safety of bioprocessing facilities
- Regulatory agency staff responsible for evaluating and overseeing the initial permitting and ongoing compliance of bioprocessing facilities with relevant process safety standards and industry guidelines.

1.4 HOW TO USE THIS GUIDELINE

Here are some suggestions on how the following chapters may be helpful to this book's readers.

CHAPTER 2 - An Overview of the Bioprocessing Industry

- For individuals new to the bioprocessing industry or new to the management system approach for managing risk, this chapter provides a description of the wide scope of operations encompassed by the industry and introduces the bioprocess lifecycle—how an idea moves from the laboratory to be manufactured and delivered to the consumer.
- For experienced employees at all levels, it presents biosafety considerations that may be outside their day-to-day involvement or bioprocess manufacturing specialty.

CHAPTER 3 - Bioprocessing Safety Management Practices

- For bioprocess safety professionals and management personnel, this chapter provides an overview of the basic parts of a bioprocessing safety management system.
- For bioprocess safety management and staff, this chapter shows the benefit from having a documented program to meet compliance needs as well as achieving operational excellence. It provides a seed outline for building or redesigning a safety management system.
- For operations management personnel, this chapter helps outline the resources necessary to implement an effective bioprocess safety management system.
- For all employees, this section provides a view of a common model for an integrated management system

CHAPTER 4 - Identifying Bioprocess Hazards

- For bioprocess safety and environmental professionals, this chapter discusses traditional personnel and environmental issues related to bioprocessing.

CHAPTER 5 - Bioprocess Design Considerations and Unit Operations

- For facility managers, bioprocess engineers and all employees, this chapter addresses the design of typical bioprocessing facilities and provides brief overviews of the basic unit operations commonly used in bioprocessing.

CHAPTER 6 - The Effects of Emerging Technology on Bioprocessing Risk Management

- For all biosafety specialists, this chapter visits the potential changes presented by new methods, new materials, new processing technology, and new biological forms.

APPENDICES

- The appendices offer a summary of large scale biosafety practices, a compilation of checklist questions for facilities, an example bioprocess characterization questionnaire, selected regulations, and references.

2
AN OVERVIEW OF THE BIOPROCESSING INDUSTRY

The bioprocessing industry combines engineering methodology with biological components to develop products for society's benefit. As technology improves and our collective knowledge of bioprocessing techniques expands, new products and more sustainable processes are relentlessly being developed. Some of the types of products or processes in use or being developed which use the metabolic processes or characteristics of living organisms include

- enzyme manufacturing,
- biochemical raw materials and nutrients (amino acids, vitamins, biopolymers, and others)
- biopharmaceuticals (such as therapeutic proteins, polysaccharides, vaccines, and diagnostics),
- semiconductor nanomaterials production,
- biofuels production,
- biofuels from algae photosynthesis,
- diagnostic markers,
- environmental remediation through the breakdown of toxic materials,
- bioreactor production of enzymes for the environmentally-friendly bleaching of wood pulp used in paper making,
- biological fabrication of semiconductor materials,
- production of fuel-cell hydrogen using bacteria to split water,
- production of fuel-cell hydrogen from carbohydrate waste materials, and
- enzyme-catalyzed conversion of vegetable oils to biodiesel.

2.1 BIOPROCESSING'S HISTORY

However exotic modern bioprocessing capabilities and potential may seem, mankind's first uses of biological processes undoubtedly had an even larger impact. Humans have been using bioprocesses for baking bread, making cheese, fermenting alcoholic beverages, and making medicines since prehistoric times.

Early people made flat bread by mixing grain meals with water and baking the resulting dough on heated rocks. Archeologists believe the Egyptians incorporated yeast to make bread about 2,600 B.C. The ancient Greeks then learned bread making from the Egyptians and later taught the method to the Romans. The Romans taught the technique to people in many parts of Europe by 200 A.D.

The earliest record of cheese making dates back to the ancient Sumerians, who are known to have made and eaten cheese as early as 3,500 B.C. The Romans experimented with cheese design, mixing sheep's and goat's milk and adding herbs and spices. The Romans also used vegetable extracts to assist with curd formation. In 59 B.C. the Roman army introduced cheese to Gaul (now France).

Beer jugs found among late Stone Age habitation sites established that intentionally fermented beverages were present at least as early as 10,000 B.C. Beer may actually have preceded bread as a staple; wine clearly appeared as a finished product in Egyptian pictographs around 4,000 BC. The production of fermented beverages spread to Greece around 2,000 B.C. and the Romans again served to spread the technology into Europe.

And lastly, approximately 5,000 years ago, moldy soybean curd was used to treat skin infections in China.

2.1.1   Bioprocessing's Historical Advancement

The timelines below note some critical developments and discoveries related to the bioprocessing industry. In every case, the development of the discovery involved applying or designing new engineering techniques—some to enhance production efficiency and some to manage risks.

2.1.1.1   **Microbiological Advancements**

- 1857, Louis Pasteur proved that yeast is a living cell that ferments sugar to alcohol.
- 1877, Pasteur showed that some bacteria kill anthrax bacilli.
- 1928, Alexander Fleming showed that growing colonies of *Penicillium notatum* inhibit *Staphylococcus* cultures.
- 1939, Florey and Chain rediscovered that Fleming's *Penicillium* could lyse bacteria, but the yield of penicillin was small and the material unstable. They realized that producing *Penicillium* on a large scale would require isolation and purification procedures that minimized product loss.
- 1942, During World War II, government incentives encouraged several pharmaceutical companies to develop cost-effective manufacturing processes for penicillin. Chemical engineers, industrial chemists, and microbiologists quickly devised methods of countercurrent extraction,

crystallization, and lyophilization to recover penicillin from the fermentation broth in an active, stable form.

2.1.1.2   Food Science and Food Process Technology Advancements

- 1957, Scientists at USDA reported the discovery of an enzyme with the ability to transform glucose to fructose (although it required arsenic as a cofactor).
- 1965, A version of this glucose isomerase enzyme that did not require arsenate was discovered in a species of *Streptomyces*.
- 1967, The first commercial shipment of high fructose corn syrup containing 42 percent fructose was sent.

Other bioprocess developments that have impacted the food processing and food additives businesses during the last 50 years included the following:

- Production of polysaccharide gums for use in foods (for example, xanthan using *Xanthomonas* bacterium)
- Production of amino acids (using *Corynebacterium*) such as L-glutamate (for monosodium glutamate)
- Production of food processing enzymes such as glucoamylase, and rennet.
- Bioprocesses for production of L-phenylalanine and L-aspartic acid for use in producing the artificial sweetener Aspartame which was approved as a food additive and for sweetening carbonated beverages during the early 1980s with broader approvals in the early 1990s.

2.1.1.3   Genetic Advancements

- 1923, Banting and Best showed that insulin from animals could be used to treat people suffering from diabetes.
- 1953, Watson and Crick showed that DNA consists of a double helix with a code of triplets of nucleotides that correspond to specific amino acids and the sequence in which they are assembled.
- 1970s, Arber, Nathans, and Smith discovered, isolated, and applied restriction endonucleases (1978 Nobel Prize in Medicine).
- 1975, Kohler and Milstein discovered how to make monoclonal antibodies to specific antigen targets in the laboratory.
- 1977, production in lab of the first human protein manufactured in a recombinant organism (*E. coli*): somatostatin.

- 1978, Separate insulin A and B chains were achieved in *E. coli* K-12, using genes synthesized and cloned for the insulin A and B chains. Eli Lilly licensed the technology and quickly developed the process, and the first recombinant DNA product, human insulin, was marketed in 1982.

- 1991, Human insulin provided an estimated 70 percent of the demand for insulin in the United States.

2.1.1.4   Future Bioprocessing Developments

As critical as early humankind's use of bioprocesses for food and medicines was to our success as a species, it will be equally critical to our future societal success to develop reliable manufacturing processes to produce the bioproducts of the future.

These new processes will almost certainly need to address the production and development of bionanotechnology, the intersection of biology and the study of controlling of matter on an atomic and molecular scale, in order to exploit the beneficial capabilities of

- stem cells,
- gene therapy,
- functional foods,
- edible vaccines,
- pharmacogenomics, and
- biosensors.

The future development of bioplastics can potentially replace many widely used hydrocarbon-based plastics. During the last 10 to 20 years there has been increased interest by a number of chemical and materials companies around the world to develop polymers that are derived entirely or partially from renewable biomass substrates using bioprocesses. Bioprocessing may also become crucial to our renewable fuel needs—from the production of ethanol to fuel cell operation.

Our lives and our surroundings will be filled with an increasingly larger percentage of materials produced either wholly or partly through bioprocessing.

2.2   INDUSTRIAL APPLICATIONS

There are five primary categories of industrial applications of bioprocessing technology:

- Pharmaceuticals
- Biochemicals
- Chemicals

- Fuels
- Food production (including nutraceuticals)

Within each of the five general categories of industrial applications, there are numerous products and associated batch or continuous process designs.

2.2.1   Processes

Bioprocesses necessarily involve the traditional safety aspects of chemical engineering, as well as the same basic unit operations:

- Fluid dynamics (including mixing)
- Heat transfer (for warming reactors, cooling materials, and removing waste heat)
- Mass transfer (for example, oxygen transfer to growing cell populations, adsorption and drying)
- Thermodynamics (refrigeration aspects of the processes)
- Mechanical processes (solids transportation and preparation)

The main categories of bioprocesses can be grouped into four types:

- Microbial fermentation
  - o Useful for antibiotics, alcohol production, bioremediation, waste treatment, and enzyme production
- Cell culture operations
  - o For example, Chinese hamster ovary (CHO) cells useful for therapeutic protein pharmaceuticals and vaccine production
- Algal culture operations
  - o For example, bioprocesses that produce biofuels and nutraceuticals
- Harvest and primary cellular recovery
  - o Many organizations consider harvest and primary cellular recovery part of purification unit operations

2.2.2   Products

One notable aspect of the bioprocessing industry is the myriad of products, from pharmaceuticals to basic chemicals and fuels, which exist today. An exciting future is possible as even more products are developed in the research laboratories. Chapter 6 – *The Effects of Emerging Technology on Bioprocess Risk Management* details some of these. A critical point to be made for all large-scale bioprocessing concerns is that it is preferable to use organisms that are considered safe.

Existing products include the following groups. Some examples are given for each.

BIOPHARMACEUTICALS
- Therapeutic proteins (including insulin and antibody drugs)
- Cellular products
- Gene therapy products
- Vaccines
- Plasma blood product derivatives
- Monoclonal antibodies

DIAGNOSTICS
- Monoclonal antibodies
- Diagnostic markers

BIOMASS PRODUCTS
- Fuels (including synthetic gas)
- Alcohols
- Glycols (polyols)
- Lactic and polylactic acid
- Organic acids
- Succinic acid
- Adhesives and resins
- Composites
- Lubricants
- Pesticides
- Fertilizers
- Plastics

2.3 THE BIOPROCESS LIFECYCLE

The description of bioprocessing's historical advancement demonstrates that the span of time between discovery to viable commercial production of a bioproduct or biopharmaceutical using that discovery can be quite lengthy. This section provides an overview of the entire bioprocessing lifecycle from discovery through decommissioning.

2.3.1   Discovery

Discovery of a new drug or biochemical product pathway typically occurs in a research laboratory setting. Typical laboratory safety conditions apply here and the quantities of materials involved are generally small.

2.3.2   Development Phase: Laboratory and Pilot Plant

Once the discovery of the metabolic, bioactive, or applied materials characteristics to be developed has occurred, activity moves from the laboratory to the pilot plant to develop the biotechnological approaches necessary for scale-up. Depending upon the end products sought, whether bioproducts or biopharmaceuticals, this phase has three common features:

- Time: The development phase can take a long time to complete (for pharmaceuticals, 11 to 15 years is common).
- Expense: This phase can be very expensive.
- High Failure Rate: There is a high attrition rate for the products developed in the laboratory. It has been estimated that less than 10% of all products (and especially biopharmaceuticals) will make it into the marketplace.

In the case of developing biopharmaceutical product expression systems, this development phase may be quite complex. There are two primary categories of expression systems:

- *In vitro* (literally in glass), taking place in a controlled environment outside of a living organism
  and
- *In vivo,* taking place within a living organism

The following list provides examples of these types of processes.

- *In Vitro* Biosynthesis
  - o   Microbial fermentation
  - o   Mammalian cell culture
  - o   Monoclonal antibody production
  - o   Insect cell culture

- *In Vivo* Biosynthesis
  - o   Transgenic dairy animals
  - o   Transgenic chickens
  - o   Transchromosomic animals
  - o   Recombinant plants

   o   Transgenic terrestrial plants

   o   Transgenic aquatic plants

2.3.3   Scale-up Phase

The scale-up phase presents special safety considerations. Generally well-known bioprocessing and chemical engineering techniques are used to increase production capacities of pilot processing equipment configurations. However, management of change of the original scaled-up equipment design becomes more important as the need to track the technical basis for making a modification may be found during test runs or initial production runs. Figure 2-1, *Typical Process Operations for a Bioprocess* below shows a basic block flow diagram.

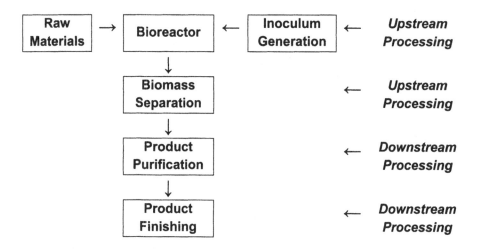

Figure 2-1   Typical Process Operations for a Bioprocess

    All aspects of traditional personnel and process safety come into play in a scale-up effort. Engineering and operations need to document the bioprocess information, analyze traditional safety and biosafety hazards, inform and train affected employees on the modifications, maintain equipment reliability, and collect monitored data to verify the desired outcomes of the scale-up effort.

As an example, the steps required to produce a recombinant *E. coli* cell line, express a recombinant protein, and complete the necessary post-biosynthesis chemical transformations and purification.

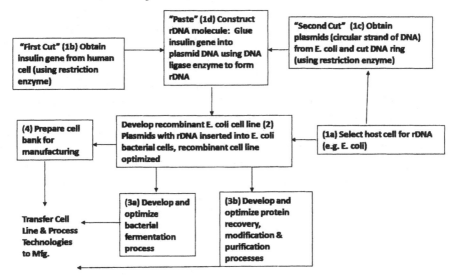

Figure 2-2   Typical Steps to Develop rDNA *E. coli*

A commercial process for producing recombinant DNA-derived human insulin is illustrated below.

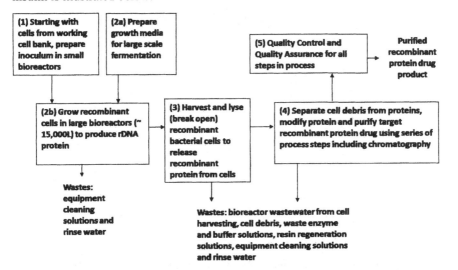

Figure 2-3   Typical Steps to Manufacture Human Insulin

2.3.4   Upstream Operations and Downstream Operations

Bioprocessing holds tremendous potential in applications such as agriculture, medicinals, fuels, enzyme production, and environmental remediation. The true economic and social impacts to these product applications have yet to be realized. At the foundation level of bioprocessing is the science of biotechnology, centered on the art and technology of fermentation and cell culture. Modern bioprocess technology is simply an extension of centuries-old techniques for developing useful products by taking advantage of naturally occurring biological activities.

The science of DNA technology and the implementation of stated engineering principles form a synergy that defines the modern biotech process. A process is defined as any operation or series of operations by which a particular objective is accomplished, such as the removal of contaminants, increased food production, or the production of a human therapeutic drug. The basic process for the manufacture of a biologic drug is shown in the figure 2-4, *Basic Biotech Manufacturing Unit Operations*.

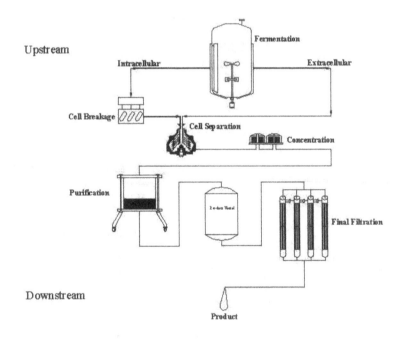

Figure 2-4   Basic Biotech Manufacturing Unit Operations

Almost every biotech-based process for the manufacture of therapeutic drug products, and many other bio-based products, can be sub-divided into two basic categories:

- Upstream processing that includes the preparation of media and the growth phase of organisms, and separation of the cell mass from the bioreactor growth media, followed by

- Downstream processing that encompasses product recovery and purification (with the product possibly being the cells, or extracellular products in solution, or components still in the cells).

Because bioprocesses use living material, they offer several advantages and challenges compared with the traditional chemical synthesis production methods. Bioprocesses usually require lower temperatures, pressures, and neutral pH for production. They use living, renewable resources as the key raw materials that support the process. These processes normally do not require a large consumption of energy to produce, but they produce a high percentage of waste materials, unrelated to the product, that must be dealt with. Keeping the organism in a state suitable for sustained product expression through the process can become quite challenging. When biohazardous cultures are involved, the challenges include protecting personnel and the environment from exposure to the product stream or wastes that have not been inactivated.

Upstream bioprocessing operations focus on cell growth by using the industrial process of fermentation as the platform. The keys are cell growth to achieve an appropriate quantity of biomass (catalyst) and production expression rate to achieve the target product concentration. The upstream process operations are designed to use selected organisms that will produce the product in large amounts, maintain stability in the culture, grow fast, and grow in the most cost-efficient (inexpensive) manner. In downstream operations, the product must be recovered from the waste cell mass for purification operations.

In the case of recombinant DNA-derived proteins or antibody substances, the target desired protein products are produced along with many other cellular co-products. Their commercialization has depended upon the development and scale-up by scientists and engineers of numerous protein recovery and purification techniques using a variety of unit operations, including synthetic membrane materials for filtration and various liquid chromatography techniques.

2.3.4.1   Inoculation / Seed and Production Biosafety Containment and Production Risk

Think big, start small. Although the details differ considerably among processes, to produce enough volume of a product, there is a defined process step called scale-up that must occur. As shown in the following figure 2-5, *The Cell Culture Scale-Up Process (Staging),* this process begins with a very small volume of material.

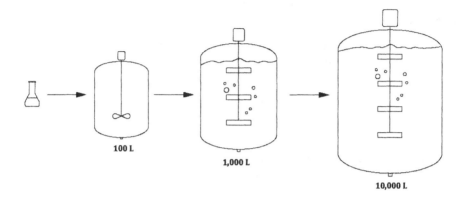

Figure 2-5   The Cell Culture Scale-Up Process (Staging)

This scale-up process begins with the transfer of a colony of genetically transformed cells to liquid media. The small volume of media, referred to as broth, allows more room for the cells to grow and provides them with more nutrients to promote the growth process. If the culture does well, it is then transferred to a spinner flask similar to the one in the following figure 2-6, *A Common Spinner Flask*. This is a commonly used vessel used for scale-up in which there is a spinner apparatus inside to keep cells suspended and aerated. Wave bags may also be used for this purpose during scale-up. Wave bags can enhance ease of validation and maintenance issues and provide reductions in energy and water use.

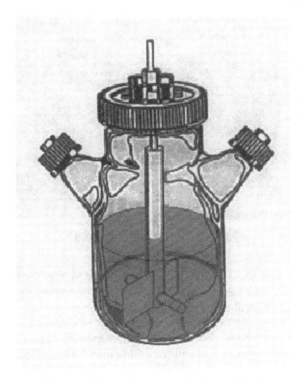

Figure 2-6   A Common Spinner Flask

Increasing the volume of broth provides more nutrients and space for the cells to grow and multiply. At some point, the volume will be such that it will become necessary to transfer the cell broth to a much larger vessel so that cell growth rate, product concentration, and product activity can be measured. This transfer step is referred to as the inoculation step of the commercial manufacturing process.

The industry is moving toward increased use of disposable technology to alleviate the need to install glass wash and autoclave operations. The reference to the spinner flask is a good example where this has been implemented in that use of smaller "wave bag" type bioreactors has replaced spinners.

The inoculation of broth from a spinner flask into a seed bioreactor (shown in figure 2-7, *Seed Bioreactors* that follows) occurs in a closed transfer. Seed train inoculum bioreactors are normally in the volume range of 25 to 50 liters.

Figure 2-7   Seed Bioreactors

The risks to successful production are potentially high at this early stage of the process. In processes based on animal cell cultures, microbial contamination is always a risk due to microbes growing much faster than animal cells. Accordingly, equipment and systems for animal cell cultures must be designed to provide fully aseptic (sterile) conditions. For microbial fermentation processes, non-host contamination is less of a risk, but still something for which appropriate equipment design is important. When manufacturing pharmaceuticals in microbial fermentation systems, for example, a sterile design and implementation is appropriate. When manufacturing products of microbial fermentation not designed for human consumption as drugs or food products, a low-bioburden equipment design may be adequate (though most plant managers will want to reduce the risk of any product loss due to bioburden contamination). Since fermentations or cell cultures will be most productive when conditions are optimized and controlled relative to the physical and nutritional needs of the production organism, effective controls of the environment (for example, temperature, pH, oxygen supply, nutrient feeding, etc.) are required. Equipment design is reflective of this point; polished surfaces that are non-reactive and easily cleaned, automation to monitor and control critical process parameters, and validated system closure are common.

Depending on the organism used, another critical aspect of the process design includes equipment design attributes focused on protection of the workers and the environment. In the United States, the National Institute of Health (NIH) and the Center for Disease Control (CDC) have developed a classification system for all microorganisms, based on their potential to harm humans and animals. Globally, the World Health Organization (WHO) provides a classification system for assigning microorganisms to various risk groups shown in Chapter 4, Table 4-1, *WHO Risk Group Classifications*. These risk group classifications are based on pathogenicity, modes of transmission, host range of the organism, effective preventative measures that can be implemented, and the availability of effective treatment. The WHO does not provide a specific listing of microorganisms into the various risk groups, since there can be differences from country to country on the relative risks the agents pose based on the endemicity of the agent in the population, the availability of suitable vectors, geophysical conditions, and other factors. So, while generally consistent, the assignment of microorganisms into the various risk groups can vary from country to country.

2.3.4.2 **Fermentation / Cell Culture**

From a unit operation perspective, "fermentation" and "cell culture" are essentially the same process. Historically, fermentation usually refers to large scale cultivation of cells or microorganisms, while cell culture typically refers to a specific type of fermentation that applies to growing cells that come from multi-cellular organisms such as animals and plants. In biotech applications, these terms are often applied to bacteria and yeast as fermentation and mammalian and insect cells as cell culture.

Fermentation is often focused on cell growth. Creating an environment that is conducive to efficient cell growth under controlled conditions is critical. The life cycle of cell growth can easily be shown in graphic form in figure 2-8, *The Cell Growth Lifecycle*.

During the phase, cells are adapting to their new environment inside the fermentor or bioreactor. As nutrients are added and the environmental conditions are developed to promote growth, the growth rate of the cells increases. During the acceleration phase, one of the key parameters that will be monitored is the doubling time or the time required for a cell to divide. For typical *E. coli* cells, this is 20 minutes; for mammalian cells it can range from 12–72 hours. This log phase of growth is critical to commercial product production to achieve the necessary volumes of material.

At the deceleration point, the apparent growth rate of the cells becomes somewhat stationary. This is where the growth rate declines quite significantly as the cells enter a state of highly reduced growth, or no significant further growth at all. This is also the level where the cell mass likely produces the largest amount of waste products, since waste production is often proportional to the total amount of cell mass present. As nutrients are used up and waste byproducts increase, growth begins to slow and then growth stops. At the end of the stationary phase, the cells will typically begin to die due to the environmental conditions present. The expression of product is often proportional to the total amount of cell mass generated. Methods exist for specifically stimulating product expression after a peak in cell mass has been achieved. These methods include the use of specific chemical inducers as well as shifts in environmental conditions.

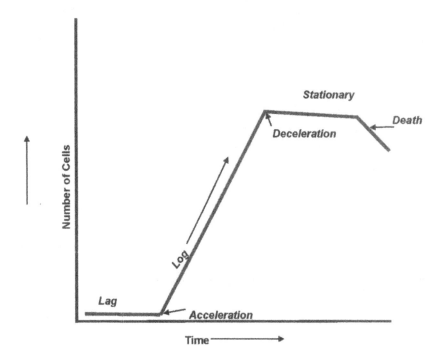

Figure 2-8   The Cell Growth Lifecycle

There are significant differences in bioprocesses that use microbial organisms compared with those that use mammalian cells. Typical characteristics of bacterial and mammalian cell lines and bioprocess parameters are shown in Table 2-1, *Comparison of Process Parameters and Characteristics of Recombinant Bacterial and Mammalian Cell Lines for Production of Recombinant DNA Protein Products*. Because cell growth rates are much different (often expressed as cell doubling times), the volumetric productivity (grams of product per liter of bioreactor per unit of time) is typically much higher for a bioprocess using a recombinant bacterial cell line compared with one using a recombinant mammalian cell line. The difference in cell line growth rates (as described by cell doubling times) means that mammalian seed train cycle times and the main bioreactor batch cycle times are much longer than for bacterial bioprocesses. This presents challenges in keeping the bioprocess free of biocontaminants. Some processes incorporate antibiotic additives to prevent growth of undesired organisms.

TABLE 2-1 Comparison of Process Parameters and Characteristics of Recombinant Bacterial and Mammalian Cell Lines for Production of Recombinant DNA Protein Products

Process Characteristic or Parameter	Recombinant *E. coli* cell line	Recombinant CHO cell line
Cell Mass Doubling Time	<1 hour	1 day or more
Media for growth	Simple	Complex
Bioreactor Process Considerations	High agitation (not very shear sensitive) Glucose feeding Cooling load and aeration needs may be high	Low agitation (shear sensitive compared tight control on, dissolved O_2, pH, and, temperature. May control dissolved CO_2 (directly or indirectly) May include bolus or continuous nutrient feeds
Batch Cycle Time following Inoculation from Seed Train	Day(s)	Weeks
Cell Mass Accumulation	High Final cell counts $=10^9$ to 10^{10} cell/ml	Low to Moderate Final cell counts $=$ 10^6 to 10^7 cells/ml
Recombinant Protein Product	Simple protein structure (no glycosylation)	More complex protein structure (glycosylated)
Product Recovery	More difficult target recombinant (protein typically retained within cell)	More straight forward target recombinant (protein is typically released from the cell into the broth)

During upstream process development, a complex set of conditions that affect cell growth, product yield and concentration of nutrients, waste, and products must be considered. Fermentation performance is impacted by a number of different properties.

- Thermodynamics: solubility of oxygen in the nutrient broth
- Transport: moving nutrients into the cells and removing waste products
- Microkinetics: cell growth and product formation
- Viscosity: media circulation and cell size

A key activity in the fermentation or cell culture process is the preparation of the cell media. The media is most often a suspension that provides the nutrients—salts, sugars, growth factors, and other material—and the environment needed for cells to survive. Without the proper media the growth process previously described will not occur.

Cells have specific dietary needs that must be met in the media.

- Carbon: Carbon-containing compounds provide the chemical energy cells need to function. Glucose is often the primary energy source. Starches or other sugars may also be used.
- Nitrogen: Cells need a nitrogen source to produce amino acids and other molecules. The sources are usually NH_3 for microbial fermentations (solid form as a salt or gaseous ammonium hydroxide in conjunction with pH control). Mammalian cells are typically fed amino acids as the significant nitrogen source.
- Oxygen and hydrogen: Cells get oxygen and hydrogen from air and water and also from the carbon and nitrogen sources.
- Phosphorus and sulfur: These are normally supplied from inorganic salts.
- Other elements: Cells also require sodium, potassium, and magnesium salts. Other metals such as zinc, iron, and manganese will also typically be present as trace elements.
- Additives: Sometimes cells can't make all the biomolecules they need from the raw materials of the broth. Additives such as amino acids, vitamins, antibiotics, and anti-fungal agents may be necessary.

Because cells deteriorate, die, and lyse when nutrient supplies are insufficient, the development of robust media formulations is extremely important. It is also important because of the potential impact that may be realized in the downstream manufacturing operations. Different types of cells require different media. Many media formulations focus on improving the fermentation process. The challenge is to develop a formulation of additives that does not create removal issues during downstream purification of the product, which will be discussed later in the chapter. The main focus of current media development operations is to develop consistency in production and performance of large batches. The other key issue is cost.

So what might a typical media formulation look like? The examples in table 2-2, *Components of Typical Growth Media* illustrate two different product platforms, microbial and mammalian, at commercial large scale manufacturing. When developing media formulations and production procedures, there are a number of issues that need to be addressed, based on the components and the operational platform.

- Mixing mistakes can cause product failure: Improper mixing of the raw materials can lead to premature cell death due to an imbalance in the nutrient levels. This is why the design of mixing systems is important even for media components.

- Sequence of additions may be critical: Just like cooking your favorite dessert recipe, the sequence of additions and their timing may have a tremendous impact on the cell growth curve.

- All media must get into the tank and be mixed: The design of the process must factor in the technology used for material addition. While highly automated and contained delivery systems are not mandatory, manual material additions can create issues of incorrect material volumes in the formulation. The handling of large quantities of raw materials often creates safety hazards such as exposure to acids or bases or simply weight-related accidents or injuries.

- Many mammalian media components are heat-sensitive and must be filter-sterilized: The nature of mammalian cells adds complexity to media preparation. If sterile filtration is required, the use of in-line sterilizing-grade 0.22 micron filters is often the preferred method. For mammalian cell processes, sterilizing-grade 0.1 micron filters are recommended due to their ability to more effectively remove potential mycoplasma contaminants.

- Standard Operating Procedures (SOPs) may be complex and must be followed: The media formulation steps can be complicated for many products. Strict adherence to SOPs is not only a regulatory requirement for human therapeutic drug products, but it is crucial to ensure proper nutrient balance and availability for the growth process.

TABLE 2-2 Components of Typical Growth Media

Microbial	**Mammalian**
• Glucose solution	• Amino acids and vitamins
• Ammonium sulfate	• Soy protein hydrolysate solution (optional)
• Protein hydrolysate (optional)	• Methotrexate (selective agent typically only in seed train)
• Growth factors	• Trace elements, including iron and zinc
• Water	• Salts
• pH control	• Glucose
• Selective agent – typically antibiotic, for recombinant cell lines	• Potassium phosphate
	• Water

2.3.4.3   **Scale of Manufacturing**

When discussing scale-up, two terms must first be defined: small scale and large scale. Depending on the product type and the process platform, these two terms can have very different meanings. For example, for commercial manufacturing of industrial enzymes, large scale may be seen as fermentation capacity of 400,000 L, where small scale would be defined as 1,000 to 2,000 L. If you are producing monoclonal antibody (MAb) products your definition of large scale would currently fall in a range of 10,000 L to 25,000 L, but if you produce vaccines, a large scale process would be seen as volumes in the 2,500 L to 5,000 L range.

By NIH definition, small scale production of recombinant organisms is defined as any volume less than 10 L, where most work is done at the developmental stage under laboratory conditions. In Canada, 10 liters is also cited, and in Japan, 20 liters is the magic number. However, one should keep in mind that the guidelines for recombinant DNA materials are designed to protect the environment and release of materials, and refer the user to other guidelines for the additional containment required for dealing with infectious agents. In the UK, their guidelines state that it is not the volume, but the intent of the work that determines the scale. The NIH / CDC, *Biosafety in Microbiological and Biomedical Laboratories* (BMBL), defines production quantities as a volume or concentration of infectious organisms considerably in excess of those used for identification and typing (CDC / NIH). They state that there is no finite volume or concentration that can be universally cited. They recommend that the laboratory director must make the assessment based on the organism, process, equipment, and facilities used. To try to understand the impact of this, consider growing 10,000 liters of *Saccharomyces cerevisiae*, versus 100 mL of *Mycobacterium tuberculosis* (a

RG3/BSL-3 bacterium) or Ebola (a RG4/BSL-4 virus). Volume is only one consideration in the assessment of the risk involved.

Without careful bioprocess engineering attention to detail, process performance may change with scale in any fermentation or biotechnology manufacturing. This is inherently due to the fact that physical processes such as heat and mass transfer are not likely linear with reactor scale. Poor mixing or different rates of oxygen transfer into the culture broth or carbon dioxide removal from the culture broth can lead to different metabolic behavior by the cells. When making recombinant proteins, these factors could also lead to effects on the characteristics of the protein.

Traditional stirred-tank fermentation or bioreactor systems have limitations on size related to the cell culture being grown. These limitations include things such as the ability to transfer oxygen into solution from the gas phase, shear stress on the cell walls due to agitator speed, fluid viscosity, and heat transfer capabilities. Inside a larger vessel, the conditions may be much more heterogeneous than seen in smaller vessels. Mixing times, shear, and mass transfer rates may vary as size of the vessel increases, as mentioned previously. Sterility issues at larger scale also become more of a concern. Larger volumes may also take longer periods of time (and longer exposure to heat) to sterilize, which may alter media composition and as a result impact cell growth.

TABLE 2-3 Examples of Very Large Scale Bioprocessing Products

Bioproducts and examples of microorganisms used	Bioreactor volume in liters
Ethanol (Yeast: *Saccharomyces cerevisiae*)	500,000
Citric acid (Fungus: *Aspergillus niger*)	200,000
Amino acid (L-lysine Bacterium: *Brevibacteruim lactofermentum*, or *Brevibacterium flavum*, or *Corynebacterium glutamicum*)	200,000
Amino acid (L-glutamate Bacterium: *Corynebacterium glutamicum*)	150,000
Antibiotics (Fungus: *Penicillium chrysogenum*)	100,000 to 200,000
Enzyme, alpha-amylase (Bacterium: *Bacillus subtilis*)	100,000

Manufacturing facilities will typically have multiple batch bioreactors with production cycles sequenced so that a batch cycle will end every day or so. Very large bioprocess plants may have from 4 to as many as 20 large bioreactors to produce a single product or for multiple product capabilities. There will be multiple "seed trains" (smaller bioreactors) to produce the inoculum for each production batch, as is discussed in later sections of this text. The production batch cycle time and desired plant throughput will determine how many of the seed trains will be required.

The industrial development of such large scale bioprocesses for production of products was made possible by the collaboration of scientists in the biological and chemical sciences, such as microbiologists and biochemists with a variety of engineering disciplines. Chemical engineers, biochemical engineers, mechanical engineers, and process automation engineers all contributed to enhancing scale-up capabilities.

At this writing large scale bioprocess facilities have not widely used plant cell culture, insect cell culture, or other non-microbial or non-mammalian cell lines. Very large volume production applications have historically been dominated by use of bacterial, yeast and fungal cell lines generally recognized as safe.

2.3.5   General Biosafety Recommendations for Large Scale Work

Although it is not the intent to review the exhaustive procedural requirements for work at each designated biosafety level here, it is important to review the basic concepts to understand the criteria for selection of the equipment and facility options. Appendix B, *Large Scale Biosafety Guidelines,* and Appendix C, *A Generic Laboratory / Large Scale Biosafety Checklist,* both provide additional detail.

- GLSP (Good Large Scale Practices) is designated for well-characterized organisms that are not pathogenic and do not produce compounds that are toxic, allergenic or biologically active (McGarrity and Hoerner, 1995). These are Risk Group 1 agents that meet the above criteria and will have been used safely over a period of time, or have been designated as safe, and will not survive or cause adverse effects in the environment (NIH, 2002:OECD, 1992). There are no specific biosafety containment requirements for GLSP facilities. Procedures are implemented in a way that does not adversely affect the health and safety of the employee by minimizing splashing, spraying and generation of aerosols.

- BL1-LS (Biosafety Level 1—Large Scale) is for nonpathogenic organisms, but can include organisms that can cause sensitization or are opportunistic pathogens. This is comprised of Risk Group 1 organisms that do not meet the criteria for work at GLSP. The goal at this level is to minimize release of viable organisms.

- BL2-LS (Biosafety Level 2—Large Scale) is used with common pathogens, that is, Risk Group 2 agents. The operations should be designed to prevent release and employee exposure to splashing, spraying and subcutaneous exposure routes.

- BL3-LS (Biosafety Level 3—Large Scale) is for Risk Group 3 agents that may be aerosol transmissible, able to spread by insect vectors, and cause serious diseases in humans and animals. Equipment and facilities used for this level must be designed to prevent employee exposure and aerosol release of the agent within the facility, and release of the agent outside the facility.

The requirements for large scale production of Risk Group 4 agents are not addressed in this document because of their highly specialized requirements and limited use.

2.3.5.1   Facility Design

The facility design and construction provide the secondary containment that protects people outside of the immediate work area, both in other parts of the facility and the community at large.

Once all of the process, agent, and environmental issues have been determined, many of the design parameters will be dictated.

It is important to keep the basic facility design criteria in mind throughout the process.

- GLSP: No special facility requirements
- BL1-LS: Facility designed to contain large spills / releases of organisms
- BL2-LS: Facility designed to contain all spills / releases of organism
- BL3-LS: Facility designed to contain all spills / releases of organisms including aerosols

2.3.5.2   Equipment Design

There are three basic types of fermentor and bioreactor designs:

- Internal mechanical agitation
- Bubble columns
- Loop reactors

The most common type of fermentation vessel used in biomanufacturing is the stirred-tank reactor. An example is shown in figure 2-9, *Stirred-Tank Bioreactor*. The equipment components are highly flexible and provide excellent gas transfer properties. They have proven design attributes that can address issues such as gas dispersion, foaming, and heat transfer.

Figure 2-9   Stirred-Tank Bioreactor

While the traditional design of this type of component uses an internal mechanical agitator to produce fluid movement, other bioreactor designs involve use of flexible walled pre-sterilized bags that are mixed by a wave motion as shown in the following figure. These bioreactors are currently sized from 10 to hundreds of liters capacity. Figure 2-10, *Large Scale Wave Bioreactor* below shows one type.

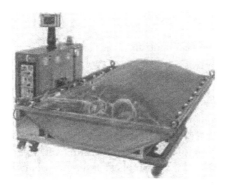

Figure 2-10   Large Scale Wave Bioreactor (Courtesy of GE Healthcare)

2.3.5.3   **Cleaning, Inactivation, and Sterilization**

These three bioprocessing steps typically expose workers to chemicals in hot solutions, steam, and high purity hot water rinses. All of these hazards may be encountered during the cleaning and sterilizing steps for vessels and piping before each new batch is begun. Traditional worker safety must be considered during these steps as well as attention to basic process safety concerns for the equipment used to perform these activities. Downstream equipment involving physical and chemical separations must also be cleaned, steam or chemically sanitized, and sterilized. (In non-sterile, bioburden reduced downstream applications, sterility may not be an absolute requirement.) This is true whether the bioprocess is a batch process, a continuous process, or a fed-batch hybrid. Equipment and media cleaning, sterilization, and inactivation must be planned and tested to ensure quality is maintained. The following list shows some aspects to consider.

- Methods to consider for cleaning, sterilization, and inactivation:
  - Clean / steam in-place methods
  - Clean / steam out-of-place methods
  - Single use (disposable) process systems
  - Total system solutions
- Equipment to consider when planning or designing cleaning, sterilization, and inactivation methods:
  - Bioreactors
  - Filtration systems
  - Centrifuges
  - Chromatography systems
  - Heat exchangers
  - Pumps

- o Tubing / piping systems
- o Valves
- o Instrumentation / controls
- o Steam supply

The term sterility assurance level (SAL) can be used to describe the term sterile as used in bioprocessing facilities. For every log reduction (10^{-1}) there is a 90% reduction in microbial population. A sterilization process that demonstrably achieves a 6-log reduction (10^{-6}) will reduce a microbial population from a million organisms (10^6) to very close to zero. Overkill cycles (typically up to 12-log reduction) are generally used for the highest assurance of sterility.

2.3.5.4   Maintenance

Maintenance activities at a large scale bioprocessing facility are an area for special attention. As with maintaining equipment that handles extremely toxic or flammable chemicals, maintaining bioprocessing equipment requires in-depth job preparation.

Pre-job analysis and planning become critical for bioprocess components. Post-job waste disposal, decontamination and inactivation issues may be unique for each type of corrective or preventive maintenance activity.

2.3.5.5   Air and Gas Emissions

Managing air emission can range from simple engineering solutions such as treating emissions for common issues like odor control (essential for microbial fermentation processes) to treating process vents with appropriate technology to maintain the barrier that prevents release of active biological agents.

Air emissions controls may involve use of high efficiency particulate air (HEPA) filters, chemical scrubbers, or thermal oxidation depending upon the specific characteristics of the bioprocess and the biosafety risk of the organism or potential contaminating organisms.

In the future, there are likely to be assessments required of carbon dioxide production and release from the largest bioprocesses, based on greenhouse gas emission regulatory requirements.

2.3.5.6   Waste Handling

Wastewater handling requirements may vary widely across the numerous bioprocesses this book addresses. This emphasizes the importance of characterizing each separate waste water stream as some processes tend to have high biological oxygen demands (BOD), high chemical oxygen demands (COD), relatively high overall nitrogen, sulfur, and phosphorous loading, or other additives such as antibiotics or metal chelating compounds. These considerations can have an impact on the operation of site or public waste water treatment systems.

In the downstream processing areas, the types and quantities of wastes will also vary widely across the different types of bioproducts and can include solvents used for liquid-liquid extraction, chemical reagent wastes from chemical transformations to convert a biological into a final product form, and aqueous chemical solutions used for regenerating chromatography resins.

Bioprocess operations typically consume significant quantities of water for both manufacturing operations and cleaning. This need for reliable water supply can challenge the hydraulic permit limits assigned to a facility.

Your facility may need to consider several issues when designing a waste handling program:

- General liquid and solid waste management
- Biohazardous / Biological waste
  - o For example, using high temperature heat exchange for continuous sterilization (inactivation) of bioprocess waste streams.
- Chemical waste
- Publicly owned treatment works (POTW) discharge restrictions
- Sanitary sewer

2.3.5.7   **Accidental Release**

All aspects of environmental release must be considered during processing, and methods of containment and remediation are necessary. Well known methods of remediation for organisms and chemical wastes need to be in place. However, there may be special controls required to prevent:

- the potential survival of harmful microorganisms,
- the establishment and dissemination of microorganisms beyond the controlled area,
- the transfer of genetic information between genetically modified organisms and other organisms, and / or
- the inherent pathogenicity of the biological materials involved.

2.3.6   **Product Safety Information**

Product safety information will have been documented and developed during the discovery and scale-up phases for the bioprocess. During the production phase, it is essential to have a complete understanding of the biosafety risks associated with the agent, including the material safety data sheets (MSDS) for the finished products and all raw materials.

Bioprocesses may have additional considerations such as the source of any animal or human material used in tissue culture or media preparation. These factors must be well understood during the manufacturing and product handling phases of operation. It is required that all organisms be carefully characterized before introduction into the laboratory, development or manufacturing plant. Refer to Appendix D, *Biological Assessment Questionnaire and Bioprocess Safety Checklist,* for an example of screening tools that can be adopted and modified to aid in the organism characterization of the risk and appropriate biosafety level assignment.

Do not confuse product safety with product contamination issues. The quality dimensions for whatever type of bioproducts a process produces must be understood. Appropriate sampling and analytical techniques need to be designed to guarantee product quality to the end user, whether a pharmaceutical product or any other type of product.

2.3.6.1   **Product Handling**

Product handling includes any type of dosage preparation, packaging, or bulk shipment preparation for the finished bioproducts. In many cases, temperature, contamination risk, and humidity can be critical controls for the products during this phase of the operation.

As discussed in Chapter 5, the facility design can be crucial for reducing the hazard exposure to both workers and the bioproduct's quality during the product handling phase.

2.3.6.2   **Material Disposal**

A bioproduct may need special stewardship considerations for the end user to understand in order to legally and safely dispose of a material.

Like bioprocess waste, finished materials and intermediates may require special handling and treatment (for example, heat, chemical neutralization, or irradiation).

2.3.6.3   **Disposable Process Technology**

The growing popularity and application of single-use bioprocessing equipment for part or all of a process step create a new material disposal concern for the industry. These costs must be built into the formula for any economic evaluation of a new bioprocess employing single-use equipment. The analysis should consider the full product life cycle as disposal technology may offset the use of chemicals, steam and other energy sources necessary for fixed equipment systems.

2.3.7   **Outsourced Manufacturing Concerns**

When considering whether to have a product or intermediate manufactured by a contract facility, there are several typical areas of their operation to evaluate.

- Facility policies, affiliations, and certifications
- Character of the area around the plant
- Personnel
- Environmental responsibility
- Security and maintenance
- Storage
- Preparedness, prevention, and emergency response
- Health and safety
- Production
- Quality assurance
- Potential exposure
- Capacity

Appendix E, *Bioprocess Facility Audit Checklist,* can be used to help assess outsourced manufacturing partners or internal facilities for general fitness for service. Customize it for the specific countries or outsourced activities you are interested in evaluating.

3
BIOPROCESSING SAFETY MANAGEMENT PRACTICES

Safety management practices for bioprocessing activities should address all aspects of the operation from laboratory to supply chain to production floor and warehousing for handling biological materials and organisms.

The overarching concept of designing and implementing management systems to maintain and improve performance is well understood. There are several models and all essentially emphasize the following root elements:

- Know what you have—both physically and in terms of information about it.
- Analyze it as necessary to understand it and all its needs.
- Operate it properly in accordance with your knowledge and analyses.
- Maintain it properly in accordance with documented methods.
- Know how you change it over time.
- Communicate the changes appropriately.
- Train your employees and key contractors on it as needed.
- Step back to look at yourself once in a while to see how you are doing.
- Develop the competencies and resources required by your business process and management system.
- Modify your business process and management system as needed to meet changing requirements.
- Communicate the status of the site's activities and findings of any annual reviews to the site senior management.

Applying this approach to bioprocess safety management works in the same way as if applying it to quality management systems, environmental management systems, process safety management systems, and traditional personnel safety management systems. This is the basic model for creating a business that can achieve sustainable development and long-term operability. Equally as important, a high quality bioprocess safety management system can help an organization develop a culture that supports safety excellence. When all employees see that the organization values and implements the system, it encourages their value of safety and environmental responsibility as well. This promotes ownership and self-policing across the organization.

Presented in this section are nine essential practices for managing bioprocess hazards. Management elements common to all safety management systems, such as emergency response, have not been included in this chapter. Figure 3-1, *Bioprocess Hazard Management Implementation Flowchart* shows how these practices come together in a logical framework for developing and maintaining bioprocess hazard management throughout the life cycle of a facility.

Some of these management practices are likely to be already in place in an existing facility required to meet quality guidelines, but they may need to be extended and applied to managing bioprocess hazards.

3.1   SAMPLE APPROACH

Many bioprocess facilities deal with chemicals and chemical hazards. These practices are compatible with chemical hazard and chemical reactivity hazard management programs as detailed in other publications, such as *Essential Practices for Managing Chemical Reactivity Hazards* published by the Center for Chemical Process Safety (CCPS). A bonus of adopting these bioprocess hazard practices will be to create a unified approach to both chemical hazard and bioprocess hazard management at a facility.

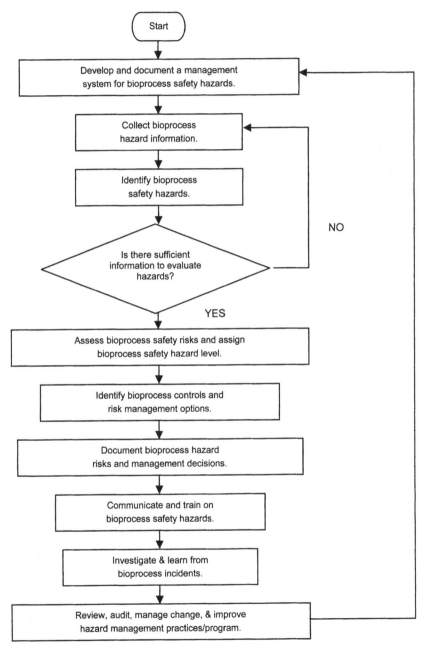

Figure 3-1   Bioprocess Hazard Management Implementation Flowchart

3.1.2   Develop and Document a System to Manage Bioprocess Safety Hazards

As the flowchart in Figure 3.1, *Bioprocess Hazard Management Implementation Flowchart* implies, managing bioprocess hazards and chemical hazards starts with a management system. To prevent incidents, a facility must not only be well designed, but also properly operated and maintained. A commitment to safety from all levels of management is essential to ensure that all safety aspects receive adequate priority. In practice, conflicts of interest may arise between safety and other goals such as production demands and budgets. In these cases, the management attitude will be decisive. In reality, such a conflict may not be an apparent one, as safety, efficiency, and product quality all depend on a reliable production facility with a low frequency of technical troubles and safety problems (CCPS 1995a).

Developing a management system is not a one-time project. It must be able to manage even subtle material, equipment or personnel changes that may have a significant effect on the safety of the operation. These may include a minor change in raw material purity, a modification to the shape of a vessel where heat transfer is important, or a change in how an operation is supervised.

3.1.3   Appoint a Biological Safety Officer

An organization should assign a biological safety officer (BSO) if it engages in large scale research or production activities involving viable organisms, particularly for those containing recombinant DNA molecules.

Depending on the size of the operation and the variety of bioprocess safety issues there may need to be studies in research, development and manufacturing. A company may establish a biosafety committee, frequently referred to as an Institutional Biosafety Committee (IBC). The IBC may also be comprised of members outside the company, such as academia or other consultants that have broad knowledge of developing issues. The institution must appoint a biological safety officer if it engages in recombinant DNA research at BL3 / BSL-3 or BL4 / BSL-4. The Biological Safety Officer's duties include, but are not be limited to

- Periodic inspections to ensure that laboratory standards are rigorously followed;
- Reporting to the organization's biosafety committee and the institution any significant problems, violations of the NIH Guidelines, and any significant research-related accidents or illnesses of which the Biological Safety Officer becomes aware unless the Biological Safety Officer determines that a report has already been filed by the Principal Investigator;

- Developing emergency plans for handling accidental spills and personnel contamination, and investigating laboratory accidents involving recombinant DNA research;
- Providing advice on laboratory security;
- Providing technical advice to Principal Investigators and the Institutional Biosafety Committee on research safety procedures;
- In the industrial setting, the Biosafety Officer coordinates the risk assessment and determines the appropriate biosafety requirements for the operation.

3.1.4 Collect Bioprocess Hazard Information

An essential practice for managing bioprocess hazards is to gather data on the bioprocess that details microbial, viral, enzyme, and contamination hazards inherent in the process and likely to be at your facility. This can be done based on inventories, reports from development laboratories, or literature citing similar bioprocesses. In each case, your management system must also include a means for detecting and evaluating any new or variant viruses or toxins brought on site for the first time.

3.1.5 Identify Bioprocess Safety Hazards

A systematic search to attempt to identify all bioprocess safety hazards, in the context of how materials will be used in the actual process, is the next step in effectively managing bioprocess safety hazards. If a particular hazard is not recognized, it is not likely to be adequately controlled. Tools used for identification include

- MSDS,
- literature surveys,
- development reports,
- laboratory data, and
- incident learning.

3.1.5.1 Point of Decision

A decision will need to be made at this point, whether sufficient information has been generated to evaluate the bioprocess safety hazards that are or will be associated with a facility. If not, then some or all of the steps outlined in the preceding sections need to be repeated before proceeding to assessing risks and identifying controls.

3.1.6   Assess Bioprocess Safety Risks and Assign Bioprocess Safety Hazard Level

To the degree that bioprocess safety hazards are identified and understood and to the extent that the facility has been designed or is currently operational, facility-specific risks can be analyzed.

Risk in this context is a combination of the likelihood and the severity of losing control of bioprocess safety hazards, taking prevention and mitigation into account.

The goal of assessing risk is to build the knowledge of bioprocess safety hazards, to understand how the hazard properties may lead to loss scenarios in the facility context, and to determine whether existing safeguards are adequate. Therefore, bioprocess safety hazard levels are assigned to assist in the assessment of risk.

Assessment of risk can be performed at any stage of facility design, development, operation or alteration. Of course, the more that is known about the bioprocess, facility, its equipment, and its operation, the more detailed the risk assessment can be.

3.1.7   Identify Bioprocess Controls and Risk Management Options

Various measures can be used to reduce the assessed risks. These measures can be classified into four types.

- Inherent (for example, DNA modified viruses that reduce the bioprocess hazard or die when containment is lost)
- Passive (for example, containment of process within vessels, piping, or vessels)
- Active (for example, instrumentation and shutdown systems)
- Procedural (for example, administrative control, PPE requirements, maintenance, emergency response, and operating procedures)

Risk control strategies in the first two categories, inherent and passive, are considered more reliable because they depend on the physical and chemical properties of the system rather than the successful operation of instruments, devices, procedures, and people. A truly inherently safer process will reduce or eliminate the hazard (Kletz 1998), rather than simply reducing its impact. These categories are not rigidly defined, and some strategies may include aspects of more than one category (Bollinger et al. 1996).

3.1.8   Document Bioprocess Safety Hazard Risks and Management Decisions

Capturing process knowledge and safety information is fundamental to many aspects of managing bioprocess safety hazards. However, merely maintaining factual design information is not sufficient. Administrative procedures and control limits are not always documented with adequate explanation of their underlying basis. Consequently, when something needs to be changed, or operating personnel need to respond to an abnormal situation, false assumptions are often made as to why controls and procedures are the way they are. Therefore, the technical basis needs to be documented, including the WHY's.

Much of the process knowledge and documentation is developed through the earlier life cycle stages of a facility. Most of these components need to be retained or kept up to date during the entire facility lifetime. Over time, ongoing training of operating personnel in process knowledge is critical to mitigate the loss of practical expertise due to personnel turnover and institutional entropy.

3.1.9   Communicate and Train on Bioprocess Safety Hazards

Communication and training are vital to the management of hazards. However, it is quite easy to pass off information as being common sense or as too obvious to require formal communication or training. Other information such as detailed bioprocess information might be considered too involved, but is often critical to achieve uniform batch to batch performance with complex biological systems. When this information relates to the control of bioprocess safety hazards, its communication, and related training should not be overlooked. All operating personnel should have a good idea of what will happen, for example, if certain materials are mixed together or if a process is operated in the wrong range.

This applies to contract workers as well as employees since all must receive the same level of safety training. How to communicate bioprocess safety hazards is not a trivial matter, especially when outside contractors are involved, since they must at a minimum, be made aware of the hazards in the work area, and appropriately protected while they are in the area. Communication with contractors and broader outsourced manufacturing issues are addressed by Early (1996) and CCPS (2000).

When previous chemical reactivity incidents have been examined, particularly where established instructions were not followed, it is often found that facility personnel did not know that violating the procedure could lead to a release. Knowledge of the bioprocess safety hazards and the adverse results from not following procedures explicitly help make procedural violations less likely.

3.1.10   Investigate & Learn from Bioprocess Incidents

During facility operation, a bioprocess incident or near miss may occur despite all efforts to effectively manage bioprocess safety hazards. An essential element of managing bioprocess safety hazards is to appropriately report and investigate every incident or near miss involving bioprocess hazards. By investing the time and effort to determine the root causes and take corrective action, as well as to document and broadly communicate the lessons learned, previously unrecognized hazards can be identified, weaknesses in the facility safeguards and management systems can be corrected, and future incidents can be avoided.

It is important to note that incident investigation techniques are essentially the same whether applied to bioprocess hazards or chemical hazards.

3.1.11 Review, Audit, Manage Change, and Improve Hazard Management Practices and Program

Managing bioprocess safety hazards to prevent releases is an ongoing endeavor, throughout the facility lifetime. Establishing a management system to ensure bioprocess safety hazards are identified and controlled is likewise not a one-time project. The management system itself, as well as the various control methods used, not only should be maintained in an operational condition but also continually improved. Program improvements should occur before lessons are learned from an incident. Avenues for maintaining and proactively improving a hazard management program include

- active monitoring,
- employee input,
- periodic reviews of program and procedures,
- audits or self assessments of various types,
- management of change, and
- keeping abreast of new technology.

Making use of all these elements is a visible sign of management commitment and an essential means of continually improving the safety management systems.

3.2 EXISTING MANAGEMENT SYSTEMS

The European Committee for Standardization (CEN) workshop agreement *CWA 15793 Laboratory biorisk management standard* is focused upon laboratory activities but is also useful as a model for bioprocess safety management overall—from laboratory to production floor. It even accounts for a behavior based safety sub-program to help encourage safety culture. The term biorisk was developed and chosen for use in this document instead of the term biosafety, because biosafety is translated as biosecurity in a number of languages (for example, bioseguridad in Spanish), and the developers wish to assure that both safety and security risks are taken into account in the management systems.

This document both references and encourages total business management system integration with the International Organization for Standardization (ISO) EN ISO 9001:2000 quality management systems; EN ISO 14001:2004 Environmental management system and OHSAS 18001:2007 occupational safety and health management systems. Whether a company chooses formal certification with ISO type systems depends upon the demands of their market. Required or not, these models promote an efficient and increasingly successful business operation.

A description of a biorisk management system using the CEN workshop agreement model follows:

BMS-ADM-01: Biorisk Management System Description
- This document provides a general description of the management system and describes its continual improvement process.

BMS-ADM-02: Biorisk Management System Policy
- This document states the organization's biorisk management policy. To quote from the agreement:
  o *The policy shall be appropriate to the nature and scale of the risk associated with the facility and associated activities and commit to:*
    - *protecting staff, contractors, visitors, community and environment from biological agents and toxins that are stored or handled within the facility;*
    - *reducing the risk of unintentional release of, or exposure to, biological agents and toxins;*
    - *reducing the risk to an acceptable level of unauthorized intentional release of hazardous biological materials, including the need to conduct risk assessments and implement the required control measures;*
    - *complying with all legal requirements applicable to the biological agents and toxins that will be handled or possessed;*
    - *with the requirements of this standard, ensuring that the need for effective biorisk management shall take precedence over all non-health and safety operational requirements;*
    - *effectively informing all employees and relevant third parties and communicating individual obligations with regard to biorisk to those groups; and*
    - *continually improving biorisk management performance.*

BMS-ADM-03: Biorisk Management Planning

- This document addresses the following items or points to more detailed documentation.
    - o Planning for hazard identification, risk assessment, and risk control
        - ▪ Planning and resources
        - ▪ Risk assessment timing and scope
        - ▪ Hazard identification
        - ▪ Risk assessment
        - ▪ Risk management
    - o Conformity and compliance
    - o Objectives, targets, and program
        - ▪ Biorisk control objectives and targets
        - ▪ Monitoring controls

BMS-ADM-04: Roles, Responsibilities, and Authorities

- This document describes the biorisk responsibilities and accountabilities for the following positions.
    - o Top management
    - o Senior management
    - o Biorisk management committee
    - o Biorisk management advisor
    - o Scientific management
    - o Occupational health
    - o Facility management
    - o Security management
    - o Animal handling personnel

BMS-ADM-05: Personnel Training, Awareness and Competence

- This document addresses the following items or points to more detailed instructions.
    - o Recruitment
    - o Competence
    - o Continuity and succession planning
    - o Training

BMS-ADM-06: Consultation and Communication

- This document addresses required and recommended communication regarding biorisk between management and employees as well as any outside stakeholders.

BMS-ADM-07: Operational Control

- This document addresses the following items or points to more detailed instructions.
  - General safety
  - Biological agents and toxin inventory and information
  - Work program, planning and capacity
  - Change management
  - Work practices, decontamination and personnel protection
    - Good microbiological technique
    - Inactivation of biological agents and toxins
    - Waste management
    - Clothing and personal protective equipment (PPE)
  - Worker health program
  - Vaccination of personnel
  - Behavioral factors and control of workers
  - Personnel reliability
  - Contractors, visitors, and suppliers
  - Exclusion
  - Infrastructure and operational management
  - Planning, design and verification
  - Commissioning and decommissioning
  - Maintenance, control, calibration, certification, and validation
  - Physical security
  - Information security
  - Control of supplies
  - Transport of biological agents and toxins
  - Personal security

BMS-ADM-08: Emergency Response and Contingency Plans

- This document addresses the following items or points to more detailed instructions or work processes. This document may need to tie into an existing integrated contingency plan (ICP) or consider other local emergency planning needs.
  - Emergency scenarios
  - Emergency plans
  - Emergency exercises and simulations
  - Contingency plans

BMS-ADM-09: Checking and Corrective Action

- This document addresses the following items or points to more detailed instructions or work processes
  - o Performance measurement and analysis of data
  - o Records, document, and data control
  - o Inventory monitoring and control
  - o Accident and incident investigation, non-conformity, corrective and preventive actions
    - Accident / incident investigation
    - Control of nonconformities
    - Corrective action
    - Preventive action
  - o Inspection and audit

BMS-ADM-10: Management Review

- This document addresses the need for an annual biorisk management review that assesses things like the following:
  - o Results of audits
  - o Compliance to SOPs and work instructions
  - o Status of risk assessment activities
  - o Status of preventive and corrective actions
  - o Follow-up actions from previous management reviews
  - o Changes that could affect the system
  - o Recommendations for improvement
  - o Results of accident / incident investigations

The ultimate goal is to engage management in assessing improvements to

- the biorisk management system,
- meeting requirements and assessing risks, and
- the resources needed to manage risk and meet organizational needs.

The CCPS recommends that users of this book use the already extensive process safety management system guidelines described in other CCPS books where appropriate. One particularly important book for most organizations is the CCPS *Guidelines for Risk Based Process Safety*.

A risk-based strategy recognizes that all hazards and risks in a facility are not equal. Using the same resource-intensive practices to manage every hazard is an inefficient use of resources. Allocating resources in a manner that focuses effort on greater hazards and higher risks allows a facility to prevent assigning an undue amount of resources to manage lower risks, thereby freeing resources for tasks that address the higher risk exposures.

Where practical, PSM system processes and practices should be integrated into an organization's overall business management system processes. Examples are the cross-functional business processes for new product development, new manufacturing process development, capital project delivery, product manufacturing, and facility maintenance planning and execution.

Table 3-1, *Comparison of a CEN-Based Biorisk Management System to OSHA PSM, EPA RMP, and CCPS Risk-Based Process Safety (RBPS)* shows how the process safety, risk management prevention program, and CCPS risk-based process safety elements may combine with biorisk management elements when a facility has both the concerns of highly hazardous chemical as well as biological hazards

The Public Health Agency of Canada's *The Laboratory Biosafety Guidelines: 3rd Edition* 2004 also offers guidance for managing biosafety.

Although focused on laboratory work, the Canadian guideline addresses production concerns as well. The agency recognizes the unique nature of each facility and organization. It asks that the following be considered when implementing a biosafety management system:

- Size of the facility (staff and square footage)
- Concentration of multiple laboratories in the facility
- Containment levels within the facility (biosafety level 2 laboratory, multiple biosafety level 3 laboratories)
- Complexity of the processes (routine diagnostics, research, large scale, recombinant work)
- Existence of shared laboratory space within the facility (multiple investigators, various organizations)
- Experimental or diagnostic animal activities within the facility (mice in containment caging, large animal housing)

TABLE 3-1 Comparison of a CEN-Based Biorisk Management System to OSHA PSM, EPA RMP, and CCPS Risk-Based Process Safety (RBPS)

BIORISK MANAGEMENT ELEMENT	PSM / RMP & RBPS ELEMENTS TO CONSIDER
Biorisk Management System Description	RMP 68.12 / 68.15 General Requirements & Management
Biorisk Management System Policy	PSM / RMP - No equivalent element RBPS - Commit to Process Safety Elements
Biorisk Management Planning	PSM / RMP - Process Safety Information PSM / RMP - Process Hazard Analysis RBPS - Understand Hazards and Risks Elements
Roles, Responsibilities, and Authorities	RMP - Risk Management Plan Management System RBPS - Commit to Process Safety Elements
Personnel Training, Awareness, and Competence	PSM / RMP - Training PSM / RMP - Mechanical Integrity (skills training) RBPS – Manage Risk Elements
Consultation and Communication	PSM - Employee Participation PSM / RMP - Management of Change PSM / RMP – Training RBPS – Manage Risk Elements
Operational Control	PSM / RMP - Operating Procedures PSM / RMP - Safe Work Practices PSM / RMP - Mechanical Integrity PSM / RMP - Contractors PSM / RMP - Management of Change RBPS - Manage Risk Elements
Emergency Response / Contingency Plans	PSM / RMP - Emergency Response RBPS - Manage Risk Elements
Checking and Corrective Action	PSM / RMP - Incident Investigation PSM / RMP - Compliance Audit RBPS - Learn from Experience Elements
Management Review	PSM / RMP - Compliance Audit RBPS - Learn from Experience Elements

Biological safety issues to be managed may include the following:

- Identifying *training* needs and assisting with the development and delivery of biosafety training programs, such as general biosafety, biosafety cabinet (BSC) use, animal biosafety, staff orientation and, for BSL-3, containment suite training
- *Performing risk assessments* when required and developing recommendations for procedural or physical modifications
- *Auditing* the effectiveness of the biosafety program and its associated management system on a regular basis
- Participating in *accident investigations* and promoting the reporting of incidents within the facility
- *Distributing new and relevant biosafety information* to staff
- *Coordinating and monitoring the decontamination, disinfection and disposal procedures* for infectious materials in the facility
- *Coordinating the receipt, shipment and transport within the facility of infectious material* according to the Workplace Hazardous Materials Information System (WHMIS) and Transportation of Dangerous Goods (TDG) regulations
- *Establishing a record keeping and secure storage system* for all infectious material entering the facility
- *Coordinating emergency response* activities
- *Maintaining liaison* with support staff, housekeeping staff, and contractors on matters related to facility biosafety

Biosafety level 3 or 4 facilities may have the additional biosafety activities.

- Certification and recertification of the laboratory
- Investigation and remediation of containment suite physical or operational failures
- Access control to the containment suites
- Liaison with applicable regulatory bodies

3.2.1   Product Stewardship for Bioproducts

The management system can also be expanded to include or reference an organization's product stewardship program. Product stewardship may also be referred to as extended product responsibility (EPR). Product stewardship is the concept that a manufacturer should know and support the best ways to safely and responsibly guide a product through the manufacturing, transportation, storage, use, and waste-handling or recycling phases of its lifecycle.

The current awareness of sustainability as an ultimate business goal promotes a realistic view of product stewardship. An organization's goal should be such that each person controlling a phase in the lifecycle of a product under stewardship—manufacturers, retailers, users, and disposers—accept personal responsibility to reduce its environmental impact. A product stewardship program should support these responsibilities.

Organizations are not able to achieve product stewardship alone. Retailers or bulk users, consumers, and the existing waste management organizations must work together to find and implement the most workable and cost-effective solutions. With the wide ranging manufacturing differences within the bioprocessing industry, each product grouping may have unique product stewardship needs.

An example of product stewardship in the bioprocessing industry is the Biotechnology Industry Organization (BIO) Product Launch Stewardship Policy. This policy, developed in consultation with the industry organization's members, addresses the specific stewardship issue of asynchronous authorizations—that is, where different countries may approve, deregulate or authorize biotech crop varieties at different times. These differences in regulations between trading partners can potentially have an adverse impact on grain product commerce.

By encouraging this stewardship approach to the introduction of biotech crops, organizations can implement a science-based policy that addresses trace amounts (adventitious presence) of biotechnology-enhanced events in raw and processed grains and oilseeds, as well as food and feed. This policy supports continued food safety for consumers, farmers, food processors, and grain handlers.

3.3   ESTABLISHING A BIOPROCESS SAFETY MANAGEMENT SYSTEM

The best approach to establishing a bioprocess safety management system is to involve the stakeholders in each element as it is developed or revised. These stakeholders can include:

- Operations management
- Maintenance management
- Research and development personnel
- Supervisory staff
- Operators and technicians
- Biological safety specialists
- Safety specialists

- Environmental specialists
- Regulatory specialists (specific to the products or processes)
- Training specialists
- Documentation specialists

The management system development team leader should bring these types of people together whenever an element of your management system affects (or is affected by) their work processes. Cooperative engagement and active analysis and discussion are the best ways to develop management processes that mesh with a facility's work processes. One goal is to make the management system as simple as possible while still meeting regulatory needs and organizational goals.

3.3.1   Select a Management System Model Based Upon Your Needs

The CEN workshop agreement CWA 15793 Laboratory biorisk management provides a realistic start to addressing essential elements. However, a facility may have special needs to consider or pre-existing management systems built to address specific regulatory or business concerns. Consider whether your biosafety management system needs to address the requirements or guidelines below.

- Occupational Safety and Health Administration (OSHA)
  - o Process Safety Management of Highly Hazardous Chemicals standard, 29 CFR 1910.119
- Occupational Health and Safety Management Systems, OHSAS 18001
- Environmental Protection Agency (EPA)
  - o Accidental Release Prevention Requirements, Risk Management Programs, Clean Air Act, Section 112 (r)(7)
- Food and Drug Administration (FDA)
- United States Department of Agriculture (USDA)
- ISO 9001:2000 quality management systems
- ISO 14001:2004 environmental management systems
- ISO 22000:2005 food safety management system
- Organizational or corporate management system requirements

Many organizations with multiple management system needs have adopted the integrated management system approach. This is essentially a total business management system based upon the understanding that true business excellence requires a system to manage five things:

- traditional product quality and customer service issues,
- personnel safety,
- process safety,

- environmental responsibility, and
- the organization's business case.

The end result is that all recognized business needs for a sustainable operation are controlled in the simplest way possible and that duplicate effort is reduced to a minimum. Many companies recognize that some of the same elements appear in safety management, quality management, process safety management, and environmental management system models. Whether or not an organization uses the integrated management system approach, a biosafety management system should interact with other pre-existing systems as needed. Often you find a biosafety management system element is adequately addressed through one of these preexisting management systems, and it merely needs to be referenced and recognized for its role in improving safety performance. An example might be the ISO or OSHA PSM requirements for training workers—your existing system may suffice.

3.3.2   Identifying the Elements that Apply to Your Operations

Identify the elements of the various applicable regulations or guidelines your stakeholders see as essential to your operation. Lay out a framework of upper level documents to capture the critical elements and analyze your existing documentation to weave the fabric of a customized approach to compliance with the system you build. Many organizations use a four-tier approach to their management system documents:

- Tier 1: Biorisk Management System Administrative Procedures
  - o From our earlier example using the CEN workshop agreement, examples of these document titles would be:
    - Biorisk Management System Description
    - Biorisk Management System Policy
    - Biorisk Management Planning
    - Roles, Responsibilities, and Authorities
    - Personnel Training, Awareness, and Competence
    - Consultation and Communication
    - Operational Control
    - Emergency Response and Contingency Plans
    - Checking and Corrective Action
    - Management Review
- Tier 2: Biorisk Management System Departmental / Divisional Procedures
  - o Where these are needed to expand on the administrative or supervisory aspects of an element, a separate document may be useful. An example would be a set of training administrative

procedures or your safe work practice procedures. Not every Tier 1 administrative procedure will necessarily need a Tier 2 (or even a Tier 3) counterpart.

- Tier 3: Work Procedures, Work Instructions, Work Processes
  - o These are the step-by-step work instructions. Examples include:
    - Operating procedures
    - Maintenance procedures
    - Laboratory procedures
    - Good manufacturing practices (cGMPs)
- Tier 4: Documentation resulting from Tiers 1, 2, and 3
  - o Tier 4 documentation is evidence that management, the staff and outside auditors can use to evaluate how well the management system is working, where there are issues, and where it can be improved. Examples include:
    - Completed forms
    - cGMPs requiring sign off by the user
    - Audits or reports required by Tier 1, 2 or 3 documents

3.3.3 Establish a Review and Approval Cycle for the Documents

The management system development team leader now needs to start identifying the right groups of stakeholders to participate in the research and draft phase of the review and approval cycle for each element.

There is a basic five-phase model for a document review and approval cycle that can be modified as needed for a specific organization. However, there are certain characteristics that should always apply.

TABLE 3-2 A Five-phase Review and Approval Cycle

Draft Phase	A writer / subject matter expert works with other subject matter experts / stakeholders to research and draft the management system steps required for a given element.
Review Phase	The document is issued to knowledgeable reviewers to obtain comments and address issues. These persons are subject matter experts and stakeholders. Consider parallel or serial review pros and cons. • It is important to include the people who do the work as well as the people who manage the work in the review phase. • In some cases the review phase may be done independently. In others, the team leader may find it useful to call a meeting with the reviewers. Management of change systems often benefit from extensive upfront communication to design a workable approach. Additional review phases can be called for if needed. • A typical review phase allows 10 working days.
Comment Incorporation Phase	In this phase, the writer resolves all the comments resulting from the previous review phase and prepares the final document for approval review.
Approval Review Phase	The document owner review occurs here. The document owner can call upon any resources he or she may need to understand and comment on the new document. At this point, it may either be approved for continuance with the next phase or re-circulated back to the writer with comments.
Publication phase	In this phase the approved document along with any required review documentation is delivered to the document control function at the facility for publication in the organization's document management system. Employee training may be planned during this phase if desired prior to approval.

3.3.4   Rolling Out the Management System to the Users

The development project team leader needs to plan for an awareness-level information session or communication to all affected employees on the new management system.

Once the entire set of Tier-1 documents has been written, anyone whose work is affected by a specific element should be provided with the tools to use the new system. For some people, the interaction with the management system is minimal. For other workers, a biorisk management element may describe a large part of their jobs and tasks.

A roll-out training or informational session typically provides

- an overview of the drivers for the new management system,
- a brief summary of the purpose and content of each new or revised document,
- specific instructions on where and how the most current documents can be accessed,
- a question and answer session on the new system, and
- refresher training on how to get a document revised if the users find a mistake or an improvement during use.

The main goal here is to enlist the users of the management system to take on the critical role of a collective conscience as they use the system. Every user should work to keep the management system viable in regard to safety and business needs. The system should be designed to encourage this goal.

3.4 BIOSAFETY TRAINING FOR THE WORKFORCE

As with any industrial or technical training analysis, it is often easy for management to simply list the things that it intuits a worker needs to know to do their job safely and efficiently. This is almost always off base by a large percentage when measured by the knowledge, skills, and attitudes a worker needs to be fully engaged and confident in his or her role in their job. The bioprocessing industry in general has an advantage; it requires extensive proceduralization for work tasks at a high level of detail. These procedures become the heart of any training system. When it comes to biosafety training, the general overview of the specific hazards at a facility is a topic covered with all new employees, and existing employees receive refresher training on a regular basis. However, encouraging workers to train on and use well-written procedures that warn of conditions in every step that might put workers or the public in harm's way when it comes to release, fire, explosion or exposure to biohazards is the best approach available. Regular training on these procedures with performance evaluation is the key to success in developing a safe and effective workforce in both the laboratory and production

All technical training benefits from the application of the instructional systems design (ISD) model for determining the specific knowledge, skills, and attitudes a worker requires to succeed in their job. The following describes a very basic approach. Compare this with how your facility designs and implements training.

Job and Task Analysis and the Instructional Systems Design Model

The analysis phase of the Instructional Systems Design (ISD) model consists of a job and task analysis based upon the equipment, operations, tools, and materials to be used as well as the knowledge and skills required for each position.

Most important in this phase is the selection of the performance and learning objectives each employee must master to be successful in their job.

Basic Steps for a Job and Task Analysis

1. First, collect information about the job position from various sources such as:
   - Job descriptions
   - Standard operating procedures, maintenance procedures, emergency response plan procedures, administrative procedures, safe work practices
   - Equipment lists
   - Operating or maintenance manuals
   - Other current data such as written reports, memos, instruction sheets, P&IDs

2. Develop a list of tasks. A task can be defined as a series of steps leading to a meaningful outcome.

3. Review the list with your Subject Matter Experts (SMEs). Subject matter experts are generally persons who can competently perform the task. This review will enable you to remove tasks that are no longer performed, add any tasks that were omitted, and correct any errors.

4. With the SMEs, rate each task in terms of frequency, importance, and difficulty.

5. Summarize the ratings to identify which tasks are critical, that is, which ones are most frequent, important, and difficult.

6. The following information is needed for each task:
   - What steps are taken to perform the task?
   - What standards are used in performing the task?
   - Under which conditions will the task be performed?
   - What tools, equipment, and references are used in performing the task?
   - What safety precautions does the person performing the task need?
   - What knowledge, skills, and attitudes are needed to perform the task?
   - What training does a person need to perform the task?

The design phase examines the best methods with which to impart the knowledge or develop the skills required to achieve the objectives. It allows management to lay out the most cost-effective training plan. Development is the phase in which the modules, such as process overview training or safety training modules, are physically constructed or purchased. As operating and maintenance procedures become training tools as well as job aids and biohazard and process safety documentation, their revision or development could be considered as part of this phase in the ISD training model. The implementation phase represents the delivery of the training. Classroom, self-study or computer-based training are typically suitable for knowledge-related objectives. Hands-on performance training, walkthrough or simulation is most appropriate for procedure tasks.

The evaluation phase of the model exists to ensure that the organization has learned from the experience of doing the training and applies that learning to the previous phases of the model. Evaluating the worker's performance on tests or during operation may lead to revisiting the task analysis to include missed tasks or possibly redistributing tasks among the positions. This phase is a self-audit of how effectively and efficiently the training was completed.

Applying this training model results in what can be seen as three levels of training topics.

1. **Fundamentals**: Fundamentals include topics such as basic biohazards, personal protective equipment and techniques, pressure, temperature, flow, general safe work practices, regulatory training, and common processing steps as appropriately indicated by the analysis phase.

2. **Process Overview**: The process overview includes topics related to the equipment configuration, chemical and physical changes, and special safe-work practices related to the operations, maintenance and materials. Emphasis should be given to any new equipment and chemical hazards the startup team will encounter

3. **Job Specific**: Job-specific topics include training on new or revised operating, safety, and maintenance procedures. It could include emergency response plan training or laboratory technician training as well if those procedures were changed.

A curriculum developed for an operator position at a facility would list
- the fundamental training the employee received upon hiring (or had completed previously),
- process overviews for each task they have performed, and
- job-specific procedures for the equipment and the batch instructions for each task they are expected to perform.

3.5 INVESTIGATING INCIDENTS

The focus of any bioprocess incident investigation must be on root cause evaluation and related corrective action to prevent recurrence of similar incidents. As described in the valuable reference *Guidelines for Investigating Chemical Process Incidents*, 2nd Edition, members of a bioprocess operating investigation team all share a common language that supports their investigation objectives efficiently and accurately. Developing a system to communicate incident investigation results and corrective actions outside of the area impacted by the incident and throughout the company is vital to improving process safety. Here are some terms and descriptions of their application that apply to all industrial incidents.

Incident—*an unusual or unexpected event, which either resulted in, or had the potential to result in*
- *serious injury to personnel,*
- *significant damage to property,*
- *adverse environmental impact, or*
- *a major interruption of process operations.*

This definition implies three categories of incidents:

- Accidents
- Near misses
- Operational interruptions

*An **accident** is an event in which property damage, detrimental environmental impact, or human loss (either injury or death) occurs.*

*A **near miss** is an event in which an accident (that is, property damage, environmental impact, or human loss) or an operational interruption could have plausibly resulted if circumstances had been slightly different.*

*An **operational interruption** is an event in which production rates or product quality is seriously impacted.*

After categorizing the incident, the next step in conducting a thorough investigation is to assemble a qualified team to determine and analyze the facts of the incident. This team's charter, using appropriate investigative techniques and methodologies, is to reveal the true underlying root causes. The terms causal factor and root cause help investigators analyze the facts and communicate with each other during the investigation phase.

*A **causal factor**, also known as a critical factor or contributing cause, is a major unplanned, unintended contributor to the incident (a negative event or undesirable condition), that if eliminated would have either prevented the occurrence, or reduced its severity or frequency.*

*A **root cause** is a fundamental, underlying, system-related reason why an incident occurred that identifies a correctable failure(s) in management systems. There is typically more than one root cause for every process safety incident.*

The third step in incident investigation is to generate a report detailing facts, findings, and recommendations. Typically, recommendations are written to reduce risk by

- improving the bioprocess technology,
- upgrading the operating or maintenance procedures or practices, and
- upgrading the management systems. (When indicated in a recommendation, this is often the most critical area.)

After the investigation is completed and the findings and recommendations are issued in the report, a system must be in place to implement those recommendations. This is the fourth step and not part of the investigation itself, but rather the follow-up related to it. It is not enough to put a technological, procedural, or administrative response into effect. The action should be monitored periodically for effectiveness and, where appropriate, modified to meet the intent of the original recommendation.

The fifth step is learning. Lessons learned from one facility's incident often have applicability to other facilities within the same organization or similar business. A management system should be in place to ensure that the understanding of the lessons learned is not restricted to a single location. A well-designed and operated incident investigation management system should set up a mechanism to communicate lessons learned to all appropriate company groups and, where appropriate, other companies with similar technology. This includes maintaining an incident log, and ensuring that incident reports become a part of the process safety information document package.

These five steps will result in the greatest positive effect when they are performed in an atmosphere of openness and trust. Management must demonstrate by both its words and actions that the primary objective is not to assign blame, but to understand what happened for the sake of preventing future incidents.

3.5.1   A Generic Procedure for Initial Biohazard Incident Response

The following steps are basic to all biohazard incidents. Compare these to your facility's approach to evaluate its status.

1. Bioworker: If exposed to biohazardous or potentially biohazardous materials, take the following steps immediately to cleanse the affected area.
    - Eyes: Rinse with water for 15 minutes.
    - Mouth: Rinse with water.

- Skin: Wash the affected area with soap and water and apply an antiseptic such as alcohol, povidone iodine, chlorhexidine, or as supplied for the specific materials being handled.
- Puncture wound: Allow the wound to bleed freely. Wash the affected area with soap and water. Apply an antiseptic such as alcohol, povidone iodine, chlorhexidine, or as supplied for the specific materials being handled.

2. Bioworker: As soon as possible, report the exposure incident to
   - the site occupational health / healthcare provider,
   - the area supervisor, and
   - plant safety personnel.

3. Bioworker: Provide each contact with any information pertaining to the material to which you were exposed, for example,
   - product name,
   - material code number, or
   - a description of the material.

4. Site occupational health / healthcare provider: If the exposure involved a specific organism, evaluate the immune status of the exposed individual. Things to consider include the following:
   - Has the individual been vaccinated?
   - Have they demonstrated a protective titer (a particular concentration of antibodies for the organism in question)?
   - What is the extent of the exposure?
   - Is the affected skin intact?
   - Is any treatment or prophylaxis necessary?

5. Site occupational health / healthcare provider: Offer the exposed individual testing as indicated if the exposure involved any human-sourced material (for example, HIV, HBV, or HCV).

   NOTE: Additional testing may be provided as appropriate, at the time of the incident and as part of the incident follow-up.

6. Site safety / biosafety personnel: As part of the data collection phase of the incident investigation, further evaluate the material to which the person was exposed and provide this information to the site occupational health / healthcare provider.

7. Site safety / biosafety personnel: If the individual was exposed to infectious or potentially infectious material, collect the following information:
   - Is the organism infectious? If so, was the material inactivated or attenuated?
   - Is the organism's antibiotic sensitivity or antiviral sensitivity known?

NOTE: For any established infectious organism used, the antimicrobial sensitivity or chemoprophylaxis should be known and readily available to the facility. The facility needs to determine whether they can maintain stocks of the treatment on site, depending on whether they have adequate medical support and the drugs have a long enough shelf life. Usually acceptable arrangements can be made with a local hospital and infectious disease specialist, when infectious agents are involved. All hospitals have programs for treating their own employees for exposure to human-sourced material, so support for those exposures are readily available.

- Does the material contain any human-sourced materials?
- Was testing performed on these materials?

NOTE: At a minimum, test for HIV antigen / NAT, Hepatitis B surface antigen (HBsAg), and HCV NAT.

- If a cell line, has the material been tested for adventitious agents?

3. Site safety / biosafety personnel: If the individual was exposed to a contaminated instrument or piece of equipment, collect the following information:
   - What was being processed on the equipment or instrument at the time of the incident?
   - What material had been previously processed on the instrument or equipment and when was that material processed?
   - Had the instrument or equipment been cleaned or decontaminated prior to the exposure?

4. Site safety / biosafety personnel: If the exposure was the result of a spill, the following additional information should be collected.
   - The liquid involved.
     o if product, which one(s)?
     o the size of the spill,
     o if waste, information regarding dilution, how long the material had been sitting since the last addition of waste, and whether any disinfectant had been added.
   - An assessment of whether appropriate clean-up procedures were followed, for example:
     o The area was evacuated as necessary.
     o Appropriate disinfectants were used for clean-up.
     o The employees involved in clean-up wore appropriate PPE.
     o Whether or not other employees were exposed.

5. Site safety / biosafety personnel: If the incident involved a contaminated sharp, the following additional information should be documented:
   - Date and time of the exposure incident

- Type and brand of sharp involved in the incident
- A description of the exposure incident including
  - the exposed employee's job classification,
  - the department or work area where the incident occurred, and
  - the task the employee was performing at the time of the incident.
  - How did the incident occur?
  - Which body part(s) were involved?
- If the sharp had engineered sharps injury protection determine following:
  - Whether the protective mechanism was activated
  - Whether the injury occurred before the protective mechanism was activated, during activation of the mechanism, or after activation of the mechanism
- If the sharp had no engineered sharps injury protection, the injured employee's opinion as to whether and how such a mechanism could have prevented the injury
- The injured employee's opinion about whether any engineering, administrative or work practice control could have prevented the injury
  - If it is determined that a safer device cannot be used, the reason that it cannot be used. (For example, plastic tubes cannot be substituted for glass since the active component sticks to plastic.)

6. Site safety / biosafety personnel: Determine whether the investigation should consider the following issues:
   - Were appropriate controls in place at the time of the incident? (That is, engineering controls such as biological safety cabinets, sharps containers, and safety cups used, PPE worn, procedures followed.)
     - If not, corrective actions must be assigned to address these deficiencies.
   - Was lack of training a root cause of the incident?
     - If determined to be a contributing factor, is there a need for additional training? (For example, a presentation to the affected departments by biosafety personnel and an analysis of courses and operating procedures, and the overall training plan to help ensure it is appropriate and complete.)
     - Assign corrective actions for training and procedures related root causes.
   - Do operating procedures need to be written or revised to prevent recurrence of future similar incidents?

7. Site safety / biosafety personnel / facility management: Follow up the incident reports corrective actions thoroughly. For example:
   - The site occupational health / healthcare provider may recommend follow-up visits for the exposed individual. Verify these are completed.

- The exposed individual should follow up with site occupational health / healthcare provider as recommended to fully address any medical issues resulting from the exposure incident. (For example, infectious agent testing, titer check, vaccination, and chemoprophylaxis.)

- Site safety personnel must document the incident in the appropriate tracking system.

- Safety management personnel should evaluate the corrective actions in the investigation report to ensure they address the root cause(s) of the incident and are adequate to prevent recurrence.

- If sharps are involved, the accident should be included in the site's annual sharps review per the requirements of the OSHA Standard on Bloodborne Pathogens (29 CFR 1910:1030).

3.6   MANAGING CHANGE

The bioprocessing industry requires rigorous management of change systems for both bioprocess safety concerns as well as quality and productivity issues.

The first step is to identify *"What is a change?"* for your process. For some facilities in the pharmaceutical industry, any difference in the steps, methodology, duration of a step, introduction of a new seed organism, or changed equipment (even an equipment replacement in kind that meets the process's original design specification) might be considered a change. For a biochemical production facility, a change might be defined as anything that is NOT a direct replacement in kind, allowing for parts replacement without strict change management. Many companies are finding that managing all work as if it was a change has certain benefits when it comes to their equipment reliability and other accounting needs for stocking and tracking equipment performance.

The second step is to define the change to a level of detail that allows experts within your organization to evaluate its personnel safety, process safety and biosafety aspects.

This multi-disciplined team's evaluation is the third step. Safety, environmental, operations, maintenance, management personnel, and other experts as needed should evaluate the change from their point of view and organizational needs. It may result in a decision to not make the planned change but most often simply defines the detailed aspects of your facility's management systems that must be modified to accommodate the change in technology, materials, methods, or equipment. Typically process documentation is widely affected—from P&IDs to GMPs and maintenance procedures.

Some organizations encourage self-inspection programs conducted by departmental staff on a more routine basis, such as weekly. This inspection would not be as inclusive as a self-audit, would take much less time and could be based on a simple checklist. It could serve as a snapshot of compliance.

3.7   REVIEWING AND AUDITING FOR CONTINUOUS IMPROVEMENT

Establishing a periodic self-evaluation cycle above and beyond any regulatory-driven audit schedule will enhance a facility's continuous improvement goals.

By looking at the elements of your management system and performing a critical self-audit, you not only find opportunities to improve safety and production issues, but also strongly reinforce the biosafety awareness of the internal audit teams.

Management has the opportunity to express that it values critical self-audits and the resulting safety and production improvements.

3.8   APPLYING BEHAVIOR-BASED SAFETY TO BIOPROCESSES

A trend among manufacturing facilities of all types is behavior-based safety. It relies upon behavioral analysis in order to effect behavior change. Behavior-based safety trains on this principle and ways to recognize and provide feedback to each other and then recruits all workers to do so whenever an unsafe behavior is observed. The foundation of this system is the fact that for every recorded accident—whether biosafety related or traditional personnel safety related—there were undoubtedly many more unsafe behaviors that did not result in a person being hurt.

As referenced from the Cambridge Center for Behavioral Studies website, these programs focus on three things:

- Environmental changes, that is, *What is it that leads to a given behavior?*
- The behavior itself, *Am I doing my job task in accordance with the organizational procedures and values?*
- And the consequences of behavior, *Were there any positive or negative responses that occurred as a result of doing that task?*

When a person, a team or an organization develops a naturally high level of operational discipline (that is, when all players consistently characterize the values of safety, environmental responsibility, product quality and enhancing organizational value in their actions), better performance is to be expected by all workers. Some benefits of well established and thoroughly monitored behavior-based safety programs at sites that display a high level of operational discipline are as follows:

- Reducing injuries and modifying employee behavior by reinforcing safe work practices and eliminating at-risk behavior
- Reducing costs related to injuries and incidents

- Developing communications skills among all workers
- Raising overall safety awareness
- Increasing observation skills
- Developing leadership skills
- Communicating the organization's commitment to safety

4
IDENTIFYING BIOPROCESS HAZARDS

4.1 KEY CONSIDERATIONS FOR ASSESSING RISK TO MANAGE BIOPROCESS SAFETY

The starting point for biological process risk assessment is knowledge about
- the organisms,
- the recombinant constructs,
- or cell lines.

Are the organisms or materials potentially infectious? Do they produce toxins, allergens, or biologically active materials? The first step is to *know what you have*. For well understood bioprocesses, this step may be a simple matter of collating information and verifying completeness. For a bioprocess operation new to a facility, there may be more involved research.

4.1.1 Testing for Bioactivity

Bioactivity is a term used for describing the beneficial or adverse effects of a compound on living matter. When the compound is a complex chemical mixture, this activity is exerted by the substance's active ingredient or pharmacophore but can be modified by the other constituents.

The main kind of biological activity is a substance's toxicity. Activity is generally dosage-dependent, and it is not uncommon to have effects ranging from beneficial to adverse for one substance when going from low to high doses. Activity depends critically on fulfillment of the absorption, distribution, metabolism, and excretion (ADME) criteria.

A material is considered bioactive if it has interaction with or effect on any cell tissue in the human body. Pharmacological activity is typically assumed to describe beneficial effects

This is a minor concern for most bioprocesses unless they are producing a biologically active compound. This would not be the most significant component of the risk assessment in most cases and does not address testing. In the case of microorganisms, the organism must be known first and testing can follow.

4.1.2   Non-biological Hazards

As identified in Chapter 1, the biological hazards at a bioprocessing facility are the primary consideration of this book—but for all facilities the traditional chemical, mechanical, and environmental hazards to workers must be considered.

4.2   BIOPROCESS RISK ASSESSMENT

In the early development stage it is essential to characterize the risks associated with the planned bioprocessing activity.

An agent-based risk assessment is the starting point for any large scale work. The scale of the work can influence the risk assessment. For example, a nonpathogenic organism that produces an extracellular toxin may not pose a problem at 40 ml, but can create significant concerns when one is dealing with 10,000 liters. Often, it is not even a real health risk that must be considered, but the potential negative publicity from an accidental release. The general public has a heightened awareness of infectious organisms thanks to the media attention paid to swine flu (H1N1), severe acute respiratory syndrome (SARS), mad cow disease, anthrax-tainted letters, *E. coli* contaminated food, and renewed threats of bioterrorism. A rational scientific analysis of the situation is generally not acceptable to the press and the public following an accidental release. This can force the institution to consider additional containment features for the facility to minimize the potential for such an incident.

4.2.1   Three Types of Assessment

Three types of assessment need to be taken into consideration: agent, process, and external environment. These considerations may not come into play for every assessment, but should be reviewed to determine their relevance to the specific situation.

4.2.2   Agent Considerations

In doing the risk assessment, a number of questions need to be answered about the organism(s) that will be used in the facility. These include, but are not limited to, the following:

- What is the highest biosafety level needed for containment of the agent(s) that will be used in the facility?

- What is the mode of transmission?

- What is the infectious dose?

- How communicable is the agent?

- Is the agent an opportunistic pathogen that could infect immunocompromised individuals?

- Is the organism a select agent (an organism that is of particular concern to the federal government because of its potential use in biological weapons), or does it have characteristics that warrant increased security and oversight?

- Has the disease that the organism causes been eradicated so that release could cause a serious public health threat by reintroducing it into the community? (for example, polio or smallpox)

- Does the agent produce any toxic, biologically active, or allergenic compounds?

- Is the agent susceptible to adventitious contaminants (that is, bacterial, fungal, mycoplasma or viral) that may be harmful to humans?

- Are vaccines, prophylaxis, or therapeutic measures available to prevent or mediate an infection?

- Is the agent endemic in the area?

- How well does the agent survive outside of the culture system?

- Can the organism transfer genetic traits to other organisms in the environment?

- How is the agent disseminated through vectors? (for example, insects)

4.2.3   **Process Considerations**

Once that information is gathered, some specific information must be put together about the process:

- Will the facility be dedicated to one agent or will a number of agents be used?

- What volume(s) of active agents will be present in the facility?

- Will the process be continuous or batch?

- Will the equipment be stationary or movable?

- What type of equipment will be used?

- What types of manipulations need to be carried out?

- Does any of the equipment or manipulations generate aerosols?

- Will the facility be required to comply with competent authority drug or device regulations, genetically modified organism (GMO) requirements, or other governmental regulations?

- What type of cleaning, disinfection, decontamination equipment, or other similar equipment is needed?

4.2.4   Environmental Considerations

The last category can roughly be termed environmental considerations. This includes questions about the local environment external to the facility.

- What are the climatic conditions in the area? (for example, temperature, humidity, and general weather conditions)

- Has dispersion modeling data been compiled?

- What is the geography of the site?

- What are the native flora and fauna?

- How close is the air supply intake or exhaust to other facilities?

- How near is the facility to private property? What is the usage of that property? (for example, industrial, school, or housing)

- Is the site security adequate for the types of organisms handled?

- How waste is treated prior to release to the local environment?

- What type of waste is released (solid / liquid) and where it is released?

4.2.5   Microorganisms

Identification of the organism is generally not an issue, since they should be well characterized by the time the effort gets to the production phase. There are several key questions to consider.

Is the organism pathogenic? If so what is the risk group? This may depend on the country. See http://www.absa.org/riskgroups/index.html for various risk group definitions and listings. Although there is some difference in the assignment of organisms to risk groups among the various lists, care must be exercised in reading the fine print, where it will often provide the same guidance. For example, Table 4-1, *WHO Risk Group Classifications*, shows the various WHO risk groups, and the Hepatitis B virus (HBV) is defined as a risk group (RG) 2 agent in the United States and a RG 3 agent in the United Kingdom. However, the US guidance defines the production / concentrated HBV should be handled at BL-3 level; while the UK guidance stipulates that specimens containing HBV can be handled at Cat 2 (BL-2) level. Therefore, each country provides the same advice, though it is presented differently.

TABLE 4-1 WHO Risk Group Classifications

RG1	**No or low individual and community risk**	A microorganism that is unlikely to cause .human disease or animal disease. These organisms are readily available in humans or the environment, but generally do not cause harm unless the person's resistance is low.
RG2	**Moderate individual risk, low community risk**	A pathogen that can cause human or animal disease but is unlikely to be a serious hazard to lab workers, the community, livestock or the environment. Lab exposure may cause serious infection, but effective prescription drugs and preventive measures are available and the risk of spread of infection is limited.
RG3	**High individual risk, low community risk**	A pathogen that usually causes serious human or animal disease but does not ordinarily spread from one infected individual to another. Effective prescription drugs and preventive measures are available.
RG4	**High individual and community risk**	A pathogen that usually causes serious human or animal disease and that can be readily transmitted from on individual to another, directly or indirectly. Effective prescription drugs and preventive measures are not usually available.

These risk groups determine the biosafety level necessary to work with the microorganisms. The definition of a biosafety level is:

A specific combination of work practices, safety equipment, and facilities which are designed to minimize the exposure of workers and the environment to infectious agents.

Is the organism an opportunistic pathogen? That is, can it infect someone who is not immunocompetent? These organisms are usually readily available in the human or environment, but generally do not cause harm unless the person's resistance is low.

Examples of opportunistic pathogens are:

- o *Candida albicans*, a normal gastrointestinal flora, a causal agent of opportunistic oral and genital infections in humans, particularly in HIV cases.
- o *Staphylococcus*, normal flora on human skin but may cause serious skin infections.
- o *Pseudomonas aeruginosa*, a common water contaminant, is a common cause of burn and external ear infections, and is the most frequent colonizer of medical devices (for example, catheters).
- o Even if the organism is not known to be infectious or an opportunistic pathogen, getting splashed in the face with a concentrated solution of organisms with a concentration of 10^{12} organisms per liter can overcome anyone's defenses. That is why some safety measures are required for large scale work with all organisms.

Is the organism allergenic? It has been shown that exposure to some organisms can stimulate an allergic response by skin contact. For example, *Staphylococcus* and some filamentous forms of yeast and molds can stimulate an allergic response. Here again, exposure to large volumes of biological proteins may cause a response—allergic or toxic—whereas a minute natural exposure would not.

- • There is a high level of concern in Europe about the allergenic potential of genetically modified organisms (GMOs).

Does the organism produce a toxin?

- • Most of the serious exotoxin producing organisms are pathogenic. For example, these might include *Clostridium botulinum*, *Vibrio cholera*, *Bacillus anthracis*, and other well-described bacteria. An exotoxin is any toxin excreted by a microorganism, including bacteria, fungi, algae, and protozoa. An exotoxin can cause damage to the host by destroying cells or disrupting normal cellular metabolism. The exotoxins are highly potent and can be life threatening.

- Endotoxins are toxins contained in the cell walls of some microorganisms, especially gram-negative bacteria (for example, *Pseudomonas* species and *E. coli*), and are released when the bacterium dies. These toxins are generally resistant to moist heat sterilization at 121°C and require dry heat treatment of at least 200°C to be broken down. They can cause fever, chills, shock, leucopenia, and a variety of other symptoms that might result depending on the particular organism and the condition of the infected person.

One should consider process conditions whereby there may be exposure to such toxins. For example, it is widely reported that wastewater treatment plant workers suffer flu-like symptoms from exposures to such toxins. By their very nature, wastewater treatment plants tend to be somewhat open and exposure may be opportunistic. Similarly, older process technologies such as plate and frame filter presses that may be used to recover products, common in enzyme production, may result in a mechanism for toxins to escape primary containment.

Is the organism a plant or animal pathogen?

When plant and animal pathogen agents are involved, it is necessary to look at what species the organism affects. If there is a risk of it affecting local flora and fauna, then stringent containment measures must be applied, and additional precautions for entry and egress of personnel may be needed.

- Such requirements are outlined in USDA / APHIS guidelines.
- In addition, there are guidelines for RDNA plants and animals in *NIH RDNA Guidelines, Appendix P and Q*.

Testing that may be required?

A special question to consider is the testing required for a microorganism. If a facility is handling infectious bacteria, it will want to know the antibiotic sensitivities of the organism so that workers and the public may be treated if there is an exposure incident. Similarly, if working with viruses, the facility will need to know if there is an effective treatment, and whether or not the virus is sensitive to it.

4.3 RECOMBINANT ORGANISMS

The risk level of a recombinant organism is based on the host organism and the vector used. Most industrial large scale processes do not involve infectious materials. In those cases where low risk organisms are involved, the determination needs to be made whether the organism meets GLSP requirements or BL1-LS. Those include:

- Host organism: Non-pathogenic, does not contain adventitious agents, and has an extended history of safe large scale use, or built-in environmental limitations that permit optimum growth in the large scale setting but limited survival without adverse consequences in the environment.
- Vector Insert: Well characterized; free from known harmful sequences; limited in size as much as possible to the DNA required to perform the intended function; does not increase the stability of the construct in the environment, unless that is the intended function; poorly mobilizable; and does not transfer any resistance markers to microorganisms not known to acquire them naturally if such acquisition could compromise the use of a drug to control disease agents in human or veterinary medicine or agriculture.

Good large scale practice is recommended for low risk organisms such as those included in *NIH Guidelines, Appendix C, Exemptions under Section III-F-6*, which have built-in environmental limitations that permit optimum growth in the large scale setting but limited survival without adverse consequences in the environment. BL1-Large Scale is recommended for large scale research or production of viable organisms containing recombinant DNA molecules that require BL-1 containment at the laboratory scale and that do not qualify for good large-scale practice applications. Physical containment requirements for large scale processes are found in

- Appendix F – *Directive 2000/54/EC of the European Parliament and of the Council* and
- Appendix G – *Comparison of Good large Scale Practice (GLSP) and Biosafety Level (BL) – Large Scale (LS) Practice* (which presents appendix K of the NIH RDNA Guidelines).

4.4 CELL CULTURE

Bioprocesses relying upon non-microbial cell culture have built-in limitations that prevent them from presenting an environmental risk since the organisms are highly unlikely to survive of a rigorously controlled environment. However, they can pose risk of containing infectious agents. In order to be used in any large-scale operations, cell lines need to undergo rigorous adventitious agent testing and characterization. This is necessary for a number of reasons, such as to

- ensure that the cell line is not contaminated so that the product will not be corrupted or lost,
- protect the workers handling the material, and
- assure that the material is safe to use.

Obviously the latter is far more critical with therapeutic products, but is often required with diagnostic preparations as well.

Key factors for risk assessment are often based upon the following:

- Source of the cells: Theoretically, the closer the source organism is to humans genetically, the higher the potential risk.
- Cell source species: Rank high to lower risk; human, primate, mammalian, avian, invertebrate
- Tissue or cell type: Rank high to lower risk
  o blood and lymph cells,
  o neural tissue, endothelium, and gut mucosa
  o epithelial cells and fibroblasts
- Culture types: Primary cell lines, continuous cell lines, including hybridomas and recombinant cells.

There are three primary safety issues with animal cell lines:

- One is that cell lines can become contaminated from the source through operators or through media and additives. Adventitious agent testing is designed to pick up most contamination by microorganisms, but can never assure 100% freedom from viruses, especially endogenous or integrated viral genomes.
  o For example, the HeLa line of cells has been used since 1951 and were assumed virus-free. It was later found that the human papilloma virus was integrated into its genome.
- The second is the tumorigenicity (oncogenicity) of the cell line. Although considered low risk, there has been a report of a lab worker who developed a tumor after a needle stick exposure to a cell line.
- And the third issue is recombinant DNA concerns—uncharacterized hazards associated with the cells and their modification.

Animal cell culture is being increasingly used for production of therapeutic reagents such as monoclonal antibodies, recombinant proteins, viral vaccines, and replication incompetent viral vectors for gene therapy. Material derived from a variety of biological processes has been associated in the past with incidents involving the transmission of infectious agents, principally viruses. Thus, virological evaluation of animal cell substrates for use in the manufacture of biologicals is essential to ensure the safety of products for pharmaceutical use.

5
BIOPROCESS DESIGN CONSIDERATIONS AND UNIT OPERATIONS

The way in which the process equipment is designed, housed, and operated is critical to biosafety. In previous chapters the risk levels and administrative aspects of managing bioprocesses have been assessed. This chapter examines their physical aspects and operations.

5.1 PHYSICAL PLANT DESIGN

When designing a bioprocess manufacturing facility, there are numerous issues to analyze.

- What are the planned production demand requirements?
- Is this a dedicated site or a multi-purpose facility?
- If the site is already selected, what is the local regulatory climate?
- Is there land available and suitable for expansion?
- How is the location suited for the supply chain and distribution system?
- What is the business climate of the area selected for the facility?
- Are we starting with a set budget and can we meet it?
- Is the facility to be owned or leased?
- What is the schedule for bringing the space online?
- What is the building type and how does the proposed layout work with respect to applicable fire and building codes? Are there rated partitions or fire control area separations? What provision must be made for containment of fire-water in biological containment areas?

- Is there a clear understanding of the process and required protocols?
- Are all process equipment selected, critical steps identified, and SOPs established?
- Are there other hazards to be considered? In accordance with applicable codes, are the allowable exempt amounts of chemicals per control zone exceeded, requiring occupancy restrictions? Is there a potential for explosion, requiring blow out panels or explosion-proof electrical components? Are highly toxic materials being used? Is there a radiation hazard? Do we need to specify low flame low smoke materials?

5.1.1   Architectural Aspects

Architectural aspects are important to consider early in the design phase. Specific planning in materials of construction, building layout, and process step flow can impact the ease with which biorisk can be managed.

5.1.1.1   Finishes and Materials

Architectural finishes and materials within the production areas of large scale biologics processing facilities can vary from stick-built interior improvements to modular panel prefabricated rooms for pharmaceutical applications. Regardless of the materials and finishes selected, surfaces must be smooth, cleanable, durable, impervious to water and resistant to chemicals. Other desired attributes for finishes are those with anti-microbial properties, and surfaces that prevent mold / mildew growth and dust / particle build-up. Many combinations of materials can be used successfully for a wide variety of reasons, and it is important to carefully consider the following issues before making selections:

- How will the space be cleaned? How often? With what chemicals / disinfectants?
- If hazardous materials (toxic, corrosive, flammable, and combustible) are used for cleaning, buffer solutions, chromatography resins, and/or chromatography operations, has the quantity in use and storage, as well as, the total volume of the tanks been considered? Are H occupancies included in the design?
- What is your budget for constructing the spaces? Is it new construction or a renovation / retrofit?
- What are the temperature, humidity, particle count, and pressurization requirements? Are all room classifications determined?
- How often will equipment be changed out or the spaces reconfigured?

Flooring must be easily cleanable, yet provide the required degree of slip-resistance to prevent accidental falls when wet. This is a fine balance; there must be enough grit to provide good grip to prevent personnel from slipping, yet not so much that mops and cleaning equipment pads are shredded during daily cleaning, creating a maintenance nightmare. Typically, a monolithic epoxy flooring system is ideal, with an integral cove base at the walls. Seamless welded sheet vinyl flooring can also be used, but it is not as durable and can easily be ripped or damaged when moving equipment.

Walls must also be easily cleanable, taking care not to create any shelves where dust may collect or hard corners that are hard to wipe down. Classic clean room design often incorporates rounded or radius edged cove floor to wall, wall to wall, and wall to hard ceiling transitions for ease of cleaning. Similarly, door and window frames and glazing are typically installed flush with the wall surface and sealed for ease of wipe down and so that there are no horizontal surfaces where dust may collect. An epoxy paint system works very well on light-gauge metal framing and gypsum-board partitions and ceilings. It provides a monolithic surface and stands up very well to cleaning regimens. The impact of the entry of water and exposure to water on construction materials should be considered for the concealed as well as the exposed surfaces. Some materials, such as gypsum board, readily support mold growth after they become wet. If water entry is concealed, the damage may not be evident on the exposed surface until it is widespread.

Whatever finishes are finally chosen, be certain to consider inherently low flame spread and smoke spread ratings to minimize the risk of fire damage or smoke spread throughout sensitive bioprocessing environments.

The principles of biocontainment are provided in biosafety guidelines such as:

• WHO Laboratory Biosafety Manual

• CDC / NIH Biosafety in Microbiological and Biomedical Laboratories (BMBL)

• EC Directive 2000/54/EC on the protection of workers from risks related to exposure to biological agents at work

5.1.1.2   **Layout Strategies**

Layout strategies for clean manufacturing facilities are rigorous engineering coordination exercises, and the best approach is always an integrated design methodology. Each piece of the puzzle must be identified before the overall finished composition can be completed. Every piece of equipment has physical dimensions, requires clear space around it for operations / maintenance, and likely has some combination of requirements for mechanical, plumbing, and electrical utility connections. Large, heavy, or sensitive equipment may require structural consideration for weight, vibration isolation, or high bay clear space. Each piece of equipment is simply a step in the process, and the finished layout is the result of accumulating engineering and procedural information, analyzing it, developing options, refining the best solution, and thoughtfully implementing the design in the field.

Form follows function. The process must be fully understood prior to beginning the layout of the facility. Each process step should be evaluated for integration with upstream operations, downstream operations, and support functions. These systems must be identified and evaluated for proper physical segregations and environmental conditions. Levels of automation and isolation for each unit operation must be established. Many newer bioprocessing facilities have been designed with building management system that control environmental conditions outside of the process (for example, HVAC) and a process automation system for the operation. This has implications on process safety as much of the process is not manually controlled. Operators interface with these processes through human / machine interface terminals (HMI) rather than through hard copies of cGMPs. Some references regarding advanced control systems include the following.

- CCPS/AICHE, Guidelines for Safe Automation of Chemical Processes, New York (1993)
- CCPS/AICHE, Guidelines for Safe and Reliable Instrumented Protective Systems, New York (2007)
- Instrumentation, Systems and Automation Society (ISA), ANSI/ISA 84.00.01-2004 (IEC61511 modified), Functional Safety: Safety Instrumented Systems for the Process Industry Sector, Research Triangle Park, NC (2004)

The concept of nested zones in bioprocessing facilities acknowledges that people, raw materials, and containers all go into making a finished product. Each of these things must be delivered to the building and go through various steps of preparation to come together at the critical processing area (such as the point of fill)—and all must be done in such a way to eliminate the possibility for contamination of the product. The concept is that each zone gets progressively cleaner and more controlled as people and materials flow from outside the building to the heart of the operation. Just as important, a clear method for the removal of the finished product, waste, and personnel from the critical processing area must be designed. These activities or areas will each require established protocols and procedures for decontamination, cleaning, and degowning as materials move from controlled areas back to uncontrolled areas. Note that the clean versus dirty concept is often used differently than one might expect in some biosafety applications. Dirty or contaminated are terms used to designate where the biohazardous material is located.

In situations where infectious organisms are inactivated, the risk associated with handling the material is significantly reduced; the level of containment can generally be reduced, so separation of inactivated material from the infectious processing into separate rooms is useful. This practice often enables a lower level of PPE, reduces the burden on the bio-waste system, and requires less rigorous procedures. In areas where biohazardous organisms are present, the room design must include provision for spill containment and decontamination anywhere that a leak may credibly occur. This includes piping through mechanical spaces to biological liquid waste inactivation systems if a leak may occur. Two strategies to eliminate leak potential in mechanical spaces are (1) to specify all welded pipe to transfer biological waste to the inactivation system with no clamped connections, valves, or instruments in the mechanical space, and (2) to locate the biological waste treatment system adjacent to the area where the waste is generated.

The following diagram illustrates the nested zone concept, with the critical processing area being the cleanest zone (such as grade A or Class 100), surrounded by a slightly less clean zone (such as grade C or Class 10,000) to serve as a buffer between non-classified space and the heart of the operation. The idea is similar to peeling away layers of an onion to get to the center.

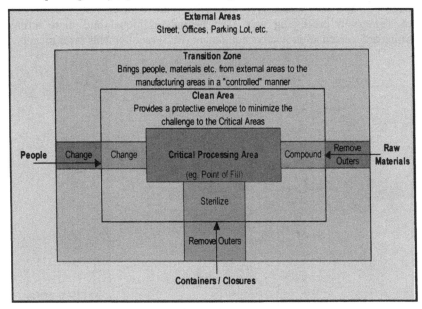

Figure 5-1   Nested Zones

Subfloors are common in areas with biocontainment requirements as an ideal place for effluent inactivation systems and fire water containment designs where required. They are also commonly used in clean room applications and referred to as subfabs for placement and staging of production floor process support equipment and other items such as cylinders of special gases to be piped to the floor above.

5.1.1.3   **People and Material Flow**

As discussed in the previous section on nested zones, people and materials all must flow into and then back out of the critical processing areas within a bioprocessing facility. This must be done in such a way that the flow of people and materials meeting certain cleanliness standards do not become contaminated because of dirty paths of travel or cross contamination. The requirements for clean manufacturing spaces will impact equipment design and personnel equipment used and brought into the clean area. This may limit the types of PPE allowed in clean areas and solutions to process safety issues in the clean spaces.

Upon arrival, personnel will have transitional areas designated for building entry and restroom / locker / change rooms for donning clean scrubs or the minimum required level of gowning prior to entering the controlled production and support areas. Once in a controlled space, people must enter clean rooms through air locks, each successively cleaner en route to the critical processing area. Each segregated processing area then having additional and more stringent gowning and personnel protective equipment (PPE) requirements prior to entry.

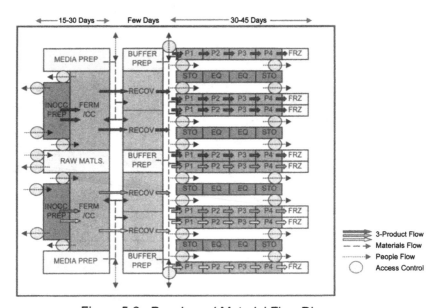

Figure 5-2   People and Material Flow Diagram

5.1.1.4   **Non-biological Hazards**

Many non-biological hazards present themselves in facility decontamination activities. Chemical, mechanical, and environmental hazards can be present.

Some companies report that most minor incidents involving processes with biosafety concerns have involved three types of hazards:

- Exposures to steam during sterilization and operating water for injection (WFI) systems
- Bulk cleaning in place (CIP) chemical handling (for example, caustic)
- Cleaning out of place (COP)

Transfer panels are commonly used to direct process and cleaning or sterilization fluids. Transfer panel connections are routinely made and broken during process operations. Check that transfer panel connections are designed with a means to identify and relieve pressure prior to breaking connections. This can be accomplished by using special connections, installing bleed ports on the jumpers, or installing a zero static valve as part of the jumper line. Installation of proximity switches on the panel to detect the presence of jumpers can also be used to confirm proper setup. Place ports that convey hazardous substances (such as steam and corrosive liquids) in an area of the panel that avoids face level and place transfer panels for maximum accessibility.

In some cases, pharma products are lyophilized, giving rise to more traditional concerns of worker exposure protection against powders. Some companies establish a biosafety level for the process and an exposure guideline or permissible exposure limit for the final product as a powder or in solution.

Routine cleaning and disinfection are generally adequate for decontaminating the facility under normal situations. However, there may be situations which a more thorough fumigation or gas decontamination of the facility is desirable. Situations that could warrant this approach would be gross contamination of the room or process, significant changes in the usage of the room or facility, major renovations or maintenance shut downs, or large equipment change-outs.

Since the gases used for fumigation have poor penetration of organic material, the first phase generally requires a thorough cleaning of the facility with a germicidal detergent. Formaldehyde vapor has historically been used for fumigation, however it poses health and safety concerns due to its toxicity. Formaldehyde gassing generally requires evacuation of the immediate area and shut-down of HVAC systems. Care must be taken using a formaldehyde gassing technique, since it can polymerize on surfaces if the temperature or humidity are not ideal and it leaves a residue that is difficult to remove and can off-gas.

Other fumigation gas options include the use of chlorine dioxide gas or vaporized hydrogen peroxide (VHP). Although extremely fast, great care must be exercised when using chlorine dioxide, since it is highly toxic. Vaporized hydrogen peroxide (also called hydrogen peroxide vapor) is also fast, but is much easier to contain and safer to use than chlorine dioxide. There are two major manufacturers of vaporized hydrogen peroxide room decontamination systems, and the requirements for use differ slightly. The general method involves

- sealing the room,
- controlling the heat and humidity to an acceptable range, and
- introducing VHP into the room for the required period of time.

The hydrogen peroxide breaks down into oxygen and water, so using a catalytic convertor to accelerate breakdown, or turning on the HVAC will return the room to safe air levels. VHP is also compatible with most materials within a room, including plastics, metals, electronics, and surface finishes. However, it may have an adverse effect on rubber over time. It can be absorbed by cellulosic materials and can off-gas. This is generally not a major issue for bioprocessing facilities.

Figure 5-3   Vaporized Hydrogen Peroxide (VHP) decontamination unit

5.1.1.5   Seismic and Building Loads

Local building codes and zoning restrictions can apply to almost any aspect of new or modified construction. Review local building codes and insurance company guidance for seismic zones and seismic restraining requirements. Additionally, these codes and standards will indicate requirements for other natural hazards such as high winds and snow loads. Equipment weights in production facilities can generate tremendous dead loads that must be considered during design of structural building systems. Large process and manufacturing equipment can routinely weigh thousands of pounds each, requiring raised floor slabs, slabs on grade, and foundation systems to be designed for very heavy dead loads to support this equipment and very heavy live loads to hold vibration within acceptable criteria. Isolation of slabs to meet extreme vibration criteria is often needed for highly calibrated and sensitive equipment. Not only must the building support these loads, but it must also be designed to withstand lateral loads and meet seismic design metrics to deal with external forces such as heavy winds or earthquakes.

Retrofitting existing structures is often required for manufacturing facilities due to the heavy equipment and building utility loads required for bioprocess operations. Most existing facilities are not designed to accommodate future loads that were never anticipated, and therefore, the existing structure must be retrofitted for the greater loads. For example, assume that an existing warehouse is chosen as a new site for manufacturing. Warehouses typically have very light requirements for utilities and roof loads, and very few are designed to have large air handlers, exhaust fans, or other large equipment placed on the roof. Placing new equipment on an existing roof will likely require additional structural support below the roof, and could require new columns and foundations to support the new loads, depending upon whether the existing structure can handle the added weight.

5.1.1.6 Hardened Construction

Certain buildings must be designed to withstand blast forces either internal or external to the building. Where hazardous materials such as flammable or combustible liquids or gases, unstable reactive materials, or other materials are used or where there is an explosion hazard damage limiting construction, explosion relief or blow out panels may be required.

5.1.1.7 Equipment Mezzanines and Subfloors

Equipment mezzanines can be a full structural floor designed to accommodate building equipment or they can be designed as an interstitial floor for equipment maintenance access only. Often, these floors are constructed as a steel structure with steel grate flooring sections or as a system of catwalks from which equipment can be adjusted, calibrated, inspected, and maintained. In many pharmaceutical and bioprocessing applications, placing equipment above the production floor can also take advantage of gravity-feed transfer of fluids and other materials between unit operations. When interstitial spaces are employed in building design, careful review of catwalks to ensure safe passage to equipment is required. This should also be reviewed when new equipment is installed in the space or the building undergoes remodeling projects. In certain instances, fall arrest anchor points may need to be considered.

From the biosafety point of view, process transfer and waste lines through mechanical spaces must be designed to prevent uncontrolled spills or the generation of airborne droplets or aerosols. Specification of all welded pipe with no clamped joints, valves or instruments throughout the mechanical space removes credible leak scenarios. Where credible leak scenarios exist, the design must include provisions to contain leaks and enable their clean-up. Procedures must be developed to protect personnel entering the area and decontaminating leaks.

Subfloors are sometimes used to house gravity-fed inactivation systems and firewater containment systems.

5.1.1.8 Heating, Ventilation, and Air Conditioning Aspects

Each bioprocessing facility requires unique consideration of its heating, ventilation, and air conditioning (HVAC) needs. Designing an efficient and effective heating, ventilation, and air conditioning system is essential to many bioprocesses.

(a) Supply and Exhaust Systems

Consider the need for 100% once-through air supply for each building or room that will be serviced. There may also be a case for the installation of emergency uninterruptible power supplies for the exhaust fans. Depending on the biosafety level and scale, the exit air as well as air supply may require HEPA filtering. Clean manufacturing facilities require more supply and exhaust air volumes that almost any other building type. Clean rooms require up to several hundred (re-circulated) air changes per hour, and most spaces where there is a chemical or biological hazard require single pass (100% outside air with no recirculation) having a minimum of 6–10 and 10-20 room air changes per hour, for laboratories and manufacturing areas, respectively. The key to practical HVAC design in a manufacturing facility is to use a closed-system design for the process manufacturing equipment where possible and use clean / containment microenvironments and isolators around equipment where open product transfer is required.

HVAC systems are generally considered to be for either *classified* or *controlled non-classified (CNC)* spaces within the production areas. Classified space refers to clean room or ISO classification, while CNC space has controlled access, but does not meet clean room requirements. There are several aspects of the HVAC system which may have an impact on the manufacturing process or product. Critical HVAC parameters that are continuously monitored and recorded for validation are typically temperature, humidity, room pressure, and directional air flow. Non-critical HVAC variables that must be monitored often include set points, duct pressure, air temperature leaving the heating or cooling coils, chilled water temperature and flow, and steam pressure. It is also critical that the HVAC design allow ease of maintenance for replacement of parts and filters, testing requirements, and the adjustment of valves and controls. Without good maintenance, the systems cannot function as designed and maintain critical parameters to ensure product safety. A clean room for biological processing is to be considered as an environment with a controlled level of particulate, expressed in the number of particles per cubic meter at a specified particle size. Table 5-1, *Clean Room Classifications* offers a comparison.

TABLE 5-1 Clean Room Classifications

ISO Class Number	Maximum number of particles/m³						Former Classification Equivalent (Federal 209E Nomenclature)
	≥0.1 µm	≥0.2 µm	≥0.3 µm	≥0.5 µm	≥1 µm	≥5 µm	
1	10	2					
2	100	24	10	4			
3	1,000	237	102	35	8		Class 1
4	10,000	2,370	1,020	352	83		Class 10
5	100,000	23,700	10,200	3,520	832	29	Class 100
6	1,000,000	237,000	102,000	35,200	8,320	293	Class 1000
7				352,000	83,200	2,930	Class 10,000
8				3,520,000	832,000	29,300	Class 100,000
9				35,200,000	8,320,000	293,000	Room air

For bioprocessing facilities, HVAC systems are designed differently for upstream vs. downstream operations. Upstream HVAC systems for areas such as fermentation are typically negative in pressure while the cleaner downstream operations, such as filling areas, are designed to maintain positive air pressure.

Exhaust air systems should be designed in such a way that HEPA filters and / or scrubbers are accessible and can be maintained safely by trained staff. Ductwork should be free of leaks to avoid contaminants being introduced back into the building or released to the atmosphere.

Consideration should also be given to the ventilation design to handle routine and non routine releases exposures to process gasses (oxygen, carbon dioxide and nitrogen) that may be used in processing. Carbon dioxide and nitrogen are asphyxiants, whereas localized concentrations of oxygen may significantly increase the risk of fire or explosion. The ventilation of areas where process materials and cultures are stored in liquid nitrogen should also be designed to ensure that a safe atmosphere is maintained. Increased air turnover rates and monitoring of gas concentrations in the area should be considered.

HEPA filtration can be used to remove almost all particulates from air streams either on the supply or exhaust air side. These filters remove 99.97% of all particles 0.22 microns in size that pass through it. Particles of this size are considered to be the most penetrating, and therefore, the most difficult to filter; particles that are either larger or smaller are filtered with even higher efficiencies. Passing air through these filters causes a significant pressure drop from one side of the filter to the other, often requiring upsized air handling and exhaust systems to achieve the required air changes.

(b) Special Exhaust Stream Mitigation

Depending upon the specific materials and biologics being handled, the exhaust stream may need to be treated before releasing to atmosphere.

Particulates can be filtered out of exhaust streams using filtration. HEPA filters and ULPA (ultra low particulate air) filters can be used to remove almost 100% of all particles, including dust, pollen, viruses, bacteria, and prions. HEPA filters with activated carbon filters are often used in areas that use radioactive materials.

Removal of fumes and gases from waste air streams is generally achieved by absorption, adsorption, or filtration. Chemical fumes are generally contained by keeping the use of chemicals within a chemical fume hood or placing a local exhaust near the source of the fumes. These fumes are typically collected within a manifolded chemical fume exhaust system where the air streams combine to dilute the number of chemical molecules in the airstream to well below safe exposure levels prior to release to the atmosphere. Care must be taken to separate incompatible waste streams that can cause dangerous chemical reactions, the most common example of which would be mixing volatile organic compounds with perchloric acid fumes, resulting in the possibility of explosion. Perchloric acid can be removed from the waste stream by passing the exhaust stream through a water scrubber as shown in figure 5-4, *A Wet Scrubber*. Other types of media can be used to filter out harmful fumes from an exhaust air stream such as activated carbon or other media that removes fumes by adsorption.

(c) HVAC Issues from a Biosafety Perspective

Directional airflow created by negative pressure differentials is used to create an air barrier between production and adjoining areas. While that is sufficient for BL1-LS, work with pathogens requires additional containment. This can be achieved in a number of ways. There are two basic designs that are most commonly used (Odum, 1995). One is an envelope system, where the internal production areas are maintained at positive pressure; and are completely surrounded by a negative pressure corridor to prevent the migration of the agent / product from the facility. This may be preferable for operations that are more vulnerable to contamination, preventing product cross contamination, and / or meeting stringent GMP requirements. In most cases, those same goals can be accomplished by using negative pressure gradients in the production area and a pressure bubble airlock, that is, air is pressurized to provide containment of hazardous material in the production area. For BL3-LS areas where more stringent containment requirements must be met and / or where there is need to prevent product cross contamination from a GMP standpoint, two adjacent airlocks can be used. The first should be a pressure bubble airlock off of the corridor, adjacent to the second, a cascading negative pressure flow airlock connected to the work area. In some cases, the corridor can serve as the first airlock. For facilities with multiple rooms, the room pressure should be most negative in the area of highest hazard, which is usually the fermentor or bioreactor. Depending on the techniques or processes involved, culture starter areas may require a similar level of containment.

The number of room air changes per hour (ACH) significantly impacts the quality of the air. In facilities that must meet Class 100,000 conditions, an ACH of 20 is not uncommon. For facilities above BL1-LS, ACH of 10–15 should be targeted. The ventilation in the rooms should be designed to maximize the air exchange in the room, generally with ceiling supply and low level returns.

The HVAC system should be sized to dissipate the heat load generated by the equipment and provide a comfortable atmosphere for employees wearing personal protective equipment.

BL1-LS facilities do not require any specialized supply or air exhaust features. Most of the following is critical for BL3-LS facilities and may be considered for BL2-LS:

- A dedicated air supply is desirable to facilitate system control and balancing.

- If the supply is shared with other areas, the use of HEPA filters at the room supply vent or airtight dampers should be used to prevent contamination of the supply system.

- Supply air should be HEPA filtered if recirculated in the facility.

- If HEPA filters are used, suitable prefilters should be used to extend the life of the HEPA.

- Ports should be provided in the filter housing to allow for periodic in-place testing of the filters.

- HEPA filtration of exhaust air should be considered depending on the agent and environmental concerns.

- Provision for testing and decontaminating HEPA filters, or use of bag in / bag out assemblies should be made.

- In certain situations, the exhaust ductwork on the contaminated side of the HEPA should be welded or sealed.

- Exhaust air from biological safety cabinets (BSC) and other containment devices should be HEPA filtered prior to discharge, preferable outside of the facility. Exhausting these devices through the room exhaust system can create problems in air balance (Ghidoni, 1999). Placing them on a separate exhaust system allows them to be used to maintain negative air flow in the facility should the room exhaust system fail.

- The exhaust and supply systems should be interlocked to prevent the facility from sustained positive pressurization.

Figure 5-4   A Wet Scrubber

(d) Microenvironments

Clean rooms are expensive to construct and maintain. An alternative to constructing the entire suite in which clean environments are required is to provide what are commonly referred to as microenvironments. Small, free-standing prefabricated clean rooms such as clean booths with hard panel walls and / or hanging plastic strips can be installed around equipment. Similarly, soft-sided clear plastic bubbles stretched over lightweight frames can provide clean environments that surround equipment or unit operations requiring clean environments. Both of these types of microenvironments can be ordered from various vendors and sized according to individual needs. They are typically provided with their own motorized fans with HEPA filtration that simply pull in room air, and then filter and recirculate clean air within the enclosure, creating a slightly positive pressure within the envelope to the surrounding room itself. Figure 5-5, *Image of a Microenvironment* provides an example.

Figure 5-5   Image of a Microenvironment

Biological safety cabinets and isolators are used to prevent the release of infectious organisms into surrounding rooms. Isolators, also commonly known as glove boxes, can be used to completely enclose any or all parts of a manufacturing line and its equipment. The isolator must be carefully designed to enable manipulations required for processing and cleaning. Except for organisms that require the highest level of containment, that is, Risk Group 4 agents and Risk Group 3 agents that are respiratory transmissible, a biological safety cabinet often provides sufficient protection for trained personnel and avoids the ergonomic issues of a glove box.

Glove boxes, as shown in figure 5-6, *A Typical Glove Box* are typically constructed of stainless steel with glass or polycarbonate walls through which a person may see and perform work within the cabinet through gasketed glove ports. Isolators may be positively or negatively pressurized to the surrounding room, depending upon needs and risk assessment for each of the particular unit operations, and they can maintain excellent particle control to achieve very clean environments. Positive pressure within the isolator can be maintained to achieve clean conditions. Alternately, negative pressure can be used within the isolator to achieve containment of chemical fumes, toxic / highly toxic compounds or infectious organisms. Almost all microenvironments can be designed to achieve class 100 clean room, pharmaceutical Grades A & B, and ISO 5 conditions to meet aseptic requirements.

Figure 5-6   A Typical Glove Box

(e) Cascading Pressure Differentials

The term directional air flow refers to the direction in which air will flow from one space to another when a door or pass-through is opened, or the direction in which infiltration or exfiltration takes place between non-air-tight spaces. This directional air flow is controlled by establishing the correct pressure differentials between spaces, and air always travels from the more positively pressurized space toward the more negatively pressurized space. Pressure differential is used to keep particles from contaminating a clean environment or to contain hazardous materials or infectious agents within a given area. Laminar flow is a term used to describe the uniform directional flow of air in a clean environment. In one type of extremely clean work area, HEPA filtered air flows directly down from the ceiling and is returned either down at floor level or directly through a perforated floor, or from the top of a microenvironment toward the bottom to knock dust or other particles straight downward.

Pressure differential diagrams are used to show room pressurizations within a floor plan. Commonly, rooms are shown with a difference from room to room of 0.05 w.g., or with cascading positive (+, ++, and so on) or negative (-, - -, and so on) designations relating to the baseline ambient environment area (0, zero) and to surrounding spaces. Air locks are required for the passage of people and / or materials into or out of clean manufacturing areas and containment suites of Biosafety Level 3 (BL-3) and above. They are also commonly found in large-scale (BL2-LS) manufacturing facilities. Air locks can be configured to provide a number of pressure differentials with respect to the room and surrounding areas to control the flow of air in a certain direction. In these suites it is essential that the process room be held at a negative pressure differential from the baseline. See Figure 5-8. The system must be designed such that under failure conditions, the airflow will not be reversed. A visual monitoring device which confirms directional air flow must be provided at the entrance. Audible alarms should be considered to notify personnel of air flow disruption. At facilities governed by GMP regulations, pressure differential alarms are routinely installed and can benefit process safety monitoring.

As an example of why cascading pressure differentials can be beneficial, consider the use of bubble-tight dampers that can be closed to facilitate any type of room decontamination (for example, when using vaporized hydrogen peroxide— VHP). The decontamination system design often includes high efficiency particulate air (HEPA) filters on exhaust air ducts from the space being decontaminated as shown in figure 5-7, *HEPA Filter Installation*.

With this configuration, a loss-of-exhaust air event from an exhaust fan failure could be mitigated through interlocks that close off supply air to prevent the processing area from becoming pressure positive with respect to exterior corridors or adjacent work areas as shown in figure 5-8, *Air Lock Types Pressure Differential Diagram*.

Figure 5-7   HEPA Filter Installation

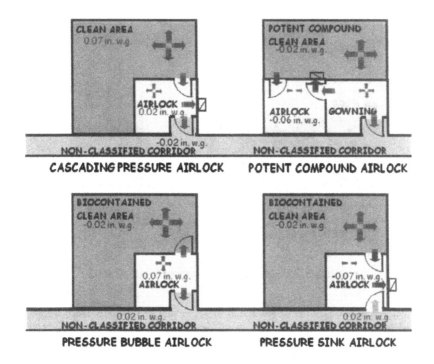

Figure 5-8   Air Lock Types Pressure Differential Diagram

(f) Containment versus Clean Room Environments

Containment is required whenever working with materials that pose a health or safety risk to workers or the public should these materials be released into the atmosphere. Examples of materials requiring containment are infectious agents such as certain viruses and bacteria, and liquids, gases or solids of a toxic, highly toxic, radioactive, unstable reactive, flammable, combustible, oxidizing, cryogenic, corrosive or poisonous nature…in short, anything posing a physical or health hazard if it were to get out in the open. All materials of these types require a risk assessment to determine the best methods of primary and secondary containment during delivery, storage, use and disposal operations. Primary containment typically refers to a vessel (that is, the fermentor or bioreactor), or closed system (that is, the biological safety cabinet, fume hood or glove box) in which the container is opened or where there is transfer of materials. Secondary containment from a biosafety standpoint generally refers to the room and associated facility attributes that prevent the material from escaping from the work area. Figure 5-9, *Full Body Positive Pressure Suit* shows a worker in an environment where hazardous materials may be present.

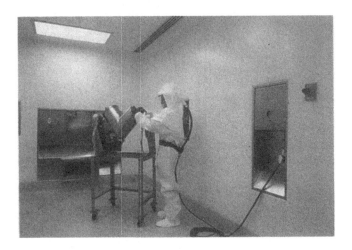

Figure 5-9   Full Body Positive Pressure Suit

Particle control is what keeps clean areas or sterile products from becoming contaminated with biological agents. Good Manufacturing Practice (GMP) requires all drugs, vaccines, and therapeutics to meet or exceed established quality assurance levels to ensure that all FDA regulated products are free from contaminants and manufactured under controlled conditions. Levels of contaminants including bioburden (that is, viable microorganisms) for all raw materials, water, and other ingredients going into the product must be within validated limits. All vessels, piping surfaces, and equipment coming in contact with the product must be clean and the atmosphere to which they are exposed must be free of particles and contaminants, including bioburden, to specified levels. These conditions must all be achieved in a manner that can be validated. For sterile products, all processing after the final sterilization must be aseptic with sterilized ingredients and equipment. Exhaust and effluent waste streams that discharge from these areas and equipment may need to be treated, filtered, inactivated or otherwise processed prior to release to the sanitary sewer system or outside atmosphere.

Conflicting regulations sometimes present design challenges in the manufacture of biologic therapeutics and vaccines. However, both containment and clean room environment requirements must be achieved. For example, polio vaccines are a derivative of the polio virus. Good manufacturing practice dictates that the product be protected from contaminants in a clean environment, often through positive pressurization within a clean environment with respect to surrounding areas. The polio virus, however, poses a serious concern over potential release of an organism that has been virtually eradicated from the world, and is only handled under rigid BSL 3 conditions. In this case, protection is often achieved through negative pressurization in the production area to provide containment of the infectious agent. In this instance, we not only need to protect the product from the people, but the people from the product. The key to these scenarios is to design a facility that achieves the containment and air quality requirements by using multiple air locks, clean / dirty corridor systems, negative envelopes, and other methods. Closed process systems provide the primary barrier by which contaminants are kept out of the product, the virus is contained, and the room envelope or microenvironment (such as an isolator) can establish and serve as the secondary containment envelope to meet biosafety requirements.

A polio vaccine made by one manufacturer is produced in a negative pressure facility which has multiple layers of positive pressure around it to achieve the required air quality. In another firm's Center for Biologics Evaluation and Research (CBER) licensed BL-3 areas, they use negative air pressure with multiple air locks and a clean / dirty corridor system that achieves containment and GMP air requirements.

5.1.1.9   Waste and Waste Treatment

Hazardous waste streams, including fire water, cannot simply be released to the sanitary sewer system. There must be a method of containing hazardous chemicals and separating them from the waste water stream prior to release. This is done through one or a combination of the following methods: mechanical separation, filtration, or neutralization. In the instance where liquids can be mechanically separated, such as with oil and water, one fluid can be skimmed off the top or drained from the bottom of a vessel to remove it. Filtration can be used in cases where the filter media or membrane traps or collects the targeted contaminants, yet allows the other fluids to flow through. In neutralization the waste stream is treated with other chemicals or materials that counteract or neutralize the properties of the waste stream such as acidity or alkalinity in order to bring the waste stream's properties to acceptable levels prior to release. In the event a waste stream cannot be treated well enough to release to the sewer, the waste must be hauled off and disposed of in some other acceptable manner in accordance with local, state, and federal regulations.

Although guidance documents for GLSP facilities have not required that the discharge of viable GLSP organisms be decontaminated, some authorities may require treatment. Regulations and biosafety guidelines require that stock cultures of all organisms are decontaminated prior to disposal. There may be additional local regulations governing specific waste parameters, for example, biological oxygen demand (BOD), level of solids, nutrients (nitrogen and phosphorus), level of pH and other analyses, which may require further processing prior to disposal.

All discharges of viable organisms and waste from BL1-LS to BL3-LS recombinant molecules require decontamination prior to disposal, as well as all wastes containing infectious organisms.

If the contaminated materials cannot be inactivated in the fermentor or bioprocessor, a decontamination tank or liquid waste treatment system may be needed. These systems are sometimes referred to as biokill systems or biowaste tanks and are used to kill or inactivate viable biological agents such as bacteria, viruses and other agents that could pose a risk to humans, animals, plants or the environment. These systems collect and treat the contaminated effluent discharge either physically or chemically. Physical treatment techniques may involve heating the effluent for a given period of time at a certain temperature, either by heating the vessel or by injecting steam. Common chemical treatments include chlorine bleach, Virkon®, or other chemical disinfectant, depending upon the organism being targeted. Another method recently developed for use in biokill systems is to heat the effluent using microwave technology.

Generally, some type of stirring or spraying system needs to be used in the biowaste treatment tank to provide adequate heat or chemical distribution, and to reduce treatment time. Waste tanks may be used to collect equipment cleaning solutions and rinses, spills, or any other potentially contaminated source. All of these factors should be considered in sizing these tanks. After treatment in the waste inactivation system, the waste stream may require further processing prior to discharge.

Inactivation processes must be validated to an appropriate level of assurance. Inactivation may be validated using the organism itself, a similar but more resistant organism, or an indicator organism that is accepted by regulatory agencies for the type of inactivation used. For example, the biological indicator for a moist heat treatment system might be *Geobacillus stearothermophillus*. It is important to note that these large-scale inactivation systems are not generally referred to as sterilization systems. This is because it is not generally practical (in terms of sampling and testing) to verify that an effluent is sterile. The inactivation systems are designed to achieve a certain reduction level of viable organism (expressed as "x log reductions" as discussed previously) in cell counts.

The time and temperature requirements for inactivation depend on the biological characteristics of the stream being treated and the safety factor desired. For example, some viruses, gram-negative bacteria or vegetative (actively growing) gram positive bacteria may be inactivated at temperatures above 70°C with exposure times of less than a minute. However other viruses and spores of gram-positive bacteria may require temperatures greater than 121°C and exposure times of minutes to hours depending on the temperature used and safety factor desired.

Production scale systems commonly use heat to inactivate cultures. Inactivation systems may be batch or continuous flow. Continuous systems heat the waste stream that is pumped through a retention loop before cooling to the temperature required for discharge. One energy conservation method is to use the effluent from the retention loop to pre-heat the inlet biological waste stream. Where steam is used to heat the system, it may be by direct steam injection or using a heat exchange surface—a vessel wall, a heating coil in a vessel, or a heat exchanger. Some considerations that favor heat exchange are:

- recovery of condensate for return to the boiler,

- heat exchange does not increase the volume of the waste stream, and

- with a continuous loop, the quality of the steam supply will not affect flow performance.

One negative consideration is that heat exchange surfaces are more vulnerable to fouling from deposition of denatured protein.

Other factors to consider with water effluent streams include the impacts of Chemical Oxygen Demand (COD), Biological Oxygen Demand (BOD), and other analytes that might have permit limits. Hydraulic flows of wastewater can be substantial where large Clean-in Place (CIP) systems are in use. In some cases, process streams may cause foaming or other disturbances in wastewater effluent. Foaming at the treatment system can cause loss of containment. The design of the system may need to consider curbing, diking, or other means to contain foaming. In short, be certain to fully characterize effluents before moving operations from small-scale to large-scale manufacturing facilities. Figure 5-10, *Effluent Decontamination System* shows a typical system.

For large scale work with pathogens, an autoclave, or other method for decontamination, should be available to process contaminated materials within the facility. The material generally poses a higher risk due to the concentration and volumes involved. A double door autoclave with access from inside and outside the facility is preferred.

Figure 5-10   Effluent Decontamination System

5.1.1.10 **Process Support Systems: High Purity Water**

USP water is an acronym for United States Pharmacopeia (grade) water. Water for Injection (WFI) is the highly purified grade of water used in the manufacturing of drugs, vaccines, and therapeutics regulated by the FDA and other world pharmaceutical industry regulatory bodies. This grade of water has very low chlorine, ionic salt, and endotoxin content and meets all USP testing requirements for pH, ammonia, chlorides, calcium, sulfates, CO_2, endotoxins, sterility, total organic carbon (TOC) content, and conductivity. This is achieved through a number of processes including filtration, reverse osmosis, deionization, and distillation. In most large-scale operations where WFI is required, it is produced on site using the local water supply as the feed water. For small volume needs, WFI can be purchased in sterile containers.

RO water is an acronym for high purity water obtained by reverse osmosis, where solutes (such as salt ions in the desalinization of sea water) are removed from the water by passing the solution through a membrane filter under pressure.

DI is the acronym for deionized water, where salt ions are removed from solutions by binding to special ion exchange resins that filter out these minerals, producing high-purity water that is similar in quality to distilled water.

Water quality standards for purified water have been established by a number of professional organizations, including the American Chemical Society (ACS), the American Society for Testing and Materials (ASTM), the National Committee for Clinical Laboratory Standards (NCCLS) which is now CLSI, and the U.S. Pharmacopeia (USP). The ASTM, NCCLS, and ISO 3696 classify purified water into Types I-III depending upon the level of purity. These organizations have similar, although not identical parameters for Type I-III water.

5.1.1.11 **Process Support Systems: Hand Washing Sinks and Personnel showers**

Hand-wash sinks are required for all work with biological materials. It is good practice and a part of routine protocol to wash one's hands prior to leaving an area in which that person may have been exposed to any chemicals, materials, or organisms that could pose a health risk. At the BL3 / BSL-3 level and above, hands-free sinks that can be operated either with wrist, elbow, foot pedals, or electronic motion sensors are required. These keep personnel from touching and contaminating the faucet shutoff valves. Hands-free sinks are also recommended for BL2 / BSL-2 facilities.

Personnel showers are not to be confused with emergency safety showers, which are only used in the event of a chemical or biohazardous exposure incident. Personnel showers are commonly used in biocontainment applications above the BL3 / BSL-3 level where workers may have the potential to be exposed to or come in contact with infectious or hazardous agents in a given area. In these applications workers must disrobe and shower in a pass-through shower as a part of routine protocols prior to exiting a given area in an effort to remove any viable organisms the person may have come in contact with while inside the containment envelope. Showers are generally provided for personnel working at BL3, particularly in production environments where long hours are spent in fairly extensive gowning, although these are generally provided for comfort and hygiene purposes.

5.1.2   Plant Siting Issues

5.1.2.1   Zoning & Permitting

Planning and zoning are commonly used by municipalities to determine the best use of land to serve the public interest and maintain property values. The cities of Cambridge, Massachusetts, and San Diego, California, have extensive biotechnology and pharmaceutical communities that have more refined zoning requirements than other municipalities.

Codes, covenants and restrictions (CC&Rs) are the set of guidelines and regulations commonly established for planned use developments. CC&Rs typically address the ratio of building square footage and size compared to the size of the property; front, side and rear yard setbacks and utility / jurisdictional right-of-ways or easements; number of parking spaces to be provided; building massing, height and aesthetic design issues, and landscaping requirements. These vary from state to state and region to region, ranging from very general guidelines subject to wide interpretation to large volumes of extremely specific and restrictive regulations depending upon the local business climate and public perceptions and attitudes regarding development in the area.

5.1.2.2   Regional Environmental Agencies and Environmental Impact Reports

Certain states have established regional agencies to deal with specific environmental issues created by local businesses and conditions that extend over multiple local and state agency jurisdictions and political boundaries. The state of California is an example of one state that has done so. Certain California agencies look at issues such as air quality and water quality on a regional basis but to date no specific requirements have been directed at the subject of process safety for bioprocessing. An example of one such agency is the South Coast Air Quality Management District (SCAQMD) which is the air pollution control agency for all of Orange County, California, and the urban portions of Los Angeles, Riverside, and San Bernardino counties. This region has serious air pollution problems and the SCAQMD was created with the concept of being committed to protecting the health of residents and reducing air pollution while remaining sensitive to businesses needs. Many other areas within California have similar agencies, including the San Francisco Bay area, Sacramento, and other areas in the Mojave Desert and the High Sierras.

Environmental Impact Reports (EIR) may also commonly be required prior to the development of a new facility in areas where the proposed new facility may

- have an impact on available utility and infrastructure resources,
- have a negative impact on endangered plant or wildlife species,
- require an industrial process that may cause environmental concerns,
- be in the flight path of an airport or military installation, and / or
- create the potential for significant physical or health hazard should an accident occur (such as a manufacturing facility for explosives, biological hazards, or highly toxic materials).

Risk assessments are often required in the EIR which outlines the processes planned in the new facility, the potential risks, the ways in which those risks can be effectively mitigated, and any other information required by authorities when addressing the proposed new development. Bioprocess risks can be reviewed by local residents and agencies as part of such an assessment process.

5.1.2.3   **Building and Site Security**

What are some typical security threats for a bioprocessing facility? They range from those applicable to any business to those specific to the bioprocessing industry:

- Severe weather
- Terrorism
- Disgruntled employees
- Theft of raw materials or products
- Contaminants (such as a mycoplasma disruption of virus growth in tissue culture)
- Improper use or release of recombinant or infectious materials
- Unplanned access to information (including GMPs)
- Unsecure access to the facility

Security should be thought of in terms of levels. The most secure areas are those deep within buildings where people must pass through more and more closely restricted layers of security the closer they get to the most restricted areas. The best systems are those that incorporate a way of recording which staff are coming and going in what areas and at what times. Card key (either swipe card or proximity reader) access systems are the most commonly used in this way. Access cards can be pre-programmed to allow or deny access to sensitive areas at any time. Other equipment, such as closed circuit television cameras (CCTV), can also be incorporated into the security system. Newer and more advanced systems have incorporated biometric security measures such as retinal and fingerprint scanners.

Security on a manufacturing site begins at the property or campus perimeter. Security fencing around the site is ideal to prevent unwanted pedestrian and vehicular access to the property, but where that is not possible, it still makes sense to restrict access to cars and trucks with the use of security gates. These can be operated either by punch code or proximity card at unmanned gate locations or opened by security guards at a guarded gate station. Visitors to the site should be registered and directed to park in designated areas where their actions can be monitored as much as possible as they park and walk to the buildings they are on site to visit.

Once inside the property, staff and visitors are commonly free to find their designated parking areas and walk unescorted to the buildings that they work in or are there to visit. The next level of security is typically access to secure buildings. Employees can normally enter buildings closest to parking areas at employee only entrances secured by card readers. Visitors must commonly sign in with security or in controlled reception areas upon entering buildings and be escorted at all times by staff within a secure building to ensure that they do not access restricted areas. Figure 5-11, *Retinal Scanning Device* shows one method of identifying cleared personnel.

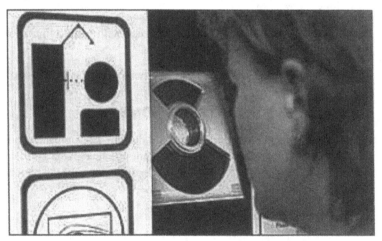

Figure 5-11   Retinal Scanning Device

Secure areas are the most sensitive and tightly controlled spaces within a building or campus setting. Typically, these areas contain narcotics, certain infectious agents, hazardous materials, or the most closely guarded intellectual property, items, and materials that require restricted access. Use of select agents and toxins require compliance with Title 42 of the Code of Federal Regulations (CFR) Parts 72 and 73 regarding the possession, use or transfer of these materials. Section 73.11 outlines security measures that must be taken in order to prevent unauthorized access, theft, loss or release of these materials. Depending upon site-specific risk analysis for each select agent present, these measures could include locked freezers and refrigerators, locked cabinets and / or controlled and monitored restricted access to the room.

When dealing with infectious materials or recombinant organisms, the biosecurity measures needed are based on the agents handled. Large scale processes have the potential for increased risk of exposure because of the volume and concentration of the agents used. However, appropriate practices, equipment and facility design can reduce the risks significantly. When select agents or other high risk agents are involved, there are specific requirements established by the licensing body, which are extensive and will not be covered here. Usually these restrictions do not apply to large-scale industrial applications, since these agents typically pose little real hazard to the public. However, any release or incident could trigger public concern and is best averted. The implementation of the basic biosecurity principles discussed above is appropriate to minimize the chance of product tampering, whether intentional or not. Specific security measures that should be considered when dealing with microorganisms include:

- Limiting access to the area, particularly only allowing access to employees who have been appropriately trained in large scale biosafety operations and requirements.

- Locking of cell bank or stock culture freezers, with further limitation of persons with access to those materials.

- Maintaining up-to-date cell / stock inventories should also be considered.

5.2 BIOPROCESS UNIT OPERATIONS

What are the primary influences that impact facility design? As companies look to develop capital assets for the manufacture of biologic products, what are the key attributes that project teams must consider? Primary influences are recognized as those listed below:

- Product attributes
  - The physical characteristics of the product being manufactured
- Process attributes
  - The materials, equipment, systems, methodologies and techniques required to deliver the product
- Facility attributes
  - The space allocations, environmental conditions, physical adjacencies, materials of construction, finishes and operating procedures required to accommodate the manufacture of the product.

Understanding these attributes leads to some fundamental concepts of facility design that are keys to successful design and operation. One of these concepts is the fact that process design is tied to facility design. The implementation of closed process systems, controlled processing capabilities, and manufacturing flexibility are the focus of many of the trends which will be discussed.

There is a recognized axiom in the world of biologic manufacturing; *the process is the product* or *the process equals the product.* This phrase defines the importance of the design and the sequencing process unit operations in relation to the attributes of the product. Changes to the process can result in unexpected or unintended changes in the product. Because biologic processes are so complex and require very precise analytical tools to ensure product safety and quality, process design must carefully consider process integration, product protection, and regulatory compliance.

Due to the nature of biological products, the fact that they are derived from living systems, product protection must also be a critical focus. To protect the product you must design systems that will succeed in product protection without adding potential contamination risk. The product must be protected by controlling the process unit operation variables, and often, the surrounding facility environment and personnel.

5.2.1   General Equipment Design Considerations

One of the fundamental foundations of bioprocess system design is the use of closed systems for production. Closed systems and equipment are used to prevent contamination of the product. They are also used as a means of containment to protect not only the product, but also the workers, the facility and the environment.

The concept of a closed processing system makes common sense. What is surprising is how closed systems are defined and referenced within the body of regulatory guidance documents and how companies differ in their implementation of a closed manufacturing concept. What is not surprising is the impact that closed systems have on facility design.

What is a closed system? For the manufacture of human therapeutic drugs, closed systems play an important role in both good engineering practice and regulatory compliance. If you go to 21 CFR Part 211 and look for a reference to closed process system or its definition, you will not find either. In fact, a search of the current literature within the industry will result in very little direct reference to defining a closed system. There are some reference materials that do begin to give definition to the concept of closed systems.

Commission Directive 2003/94/EC of the European Medicines Agency (EMEA) uses the term *closed system* in reference to equipment and system design for production. ICH Q7A of the International Conference on Harmonisation of Technical Requirements for Registration of Pharmaceuticals for Human Use (ICH) also references to the use of closed or contained systems in section IV (Buildings and Facilities) and section V (Process Equipment). However, neither of these guidance documents provides a definition for a closed system.

The ISPE Baseline Pharmaceutical Engineering Guide® for Biopharmaceutical Manufacturing Facilities does provide the following definition of a closed process or system: *"A process step (or system) that uses processing equipment in which the product is not exposed to the immediate room environment."*

The Guide goes on to state that: *"It is the Manufacturer's responsibility to define and prove closure for a process step."* So closed systems are those systems that

- use processing equipment that does not expose product to the external environment,
- allow materials to enter or leave the system via predetermined control points, and
- can be validated to prevent unwarranted material exchange between the process and the environment.

It becomes apparent that the goal is to prevent escape of the product and prevent entry of contaminants from the external environment into the product. Sounds simple, but how do you do it?

5.2.2 Closed-System Design

What are the elements of a closed-system design? How do you ensure that your equipment and components are in fact, closed? What options do you have to implement your definition of closed processing?

The American Society of Mechanical Engineers (ASME) provides a comprehensive set of guidelines applicable to closed bioprocessing systems and equipment. The critical elements of the design standards covered in these guidelines center around sterility and cleanability; each component is required to be cleaned so that the viable microbial level stays within an acceptable level, and systems must be designed to be operated in a manner that maintains product purity and cleanliness within validated boundaries. Closed aseptic systems are sterilized before processing, and those that are used to culture biological organisms maintain the culture's purity throughout processing. The accepted practices defined in the guidelines meet the intent of the GMPs for equipment design and construction. These include accepted practices for cleanability, materials of construction, connections and fittings, design for system drainability, seals and gaskets, and containment.

It is important to note that a closed system in accordance with these guidelines will not always meet the closed system definition provided in the fire and building code. To protect product cleanliness, tanks are not typically vented to the building exterior. Tank vents are filtered to protect the system boundary. The filters can become wetted and impede unobstructed venting of the process. Companies typically perform risk assessments to determine the safe location for vent terminations. As most all process vents are filtered to maintain the containment barrier and prevent ingress of contaminants, most fermenter and bioreactor vents will terminate outdoors as indicated by inherent odors, process gas use, or other driver. Conversely, where exposure risks to personnel are minimized, vents are usually terminated indoors to maintain product cleanliness. This is often the case for downstream purification operations).

Bioreactors used for the growth of microorganisms, referred to here as fermentors, and those used for cell culture can share many common attributes. In order to maintain the integrity of the culture, the culture vessel must provide an appropriate level of containment. The vessel must be constructed to be able to withstand rigorous cleaning and decontamination procedures (Bailey and Ollis, 1977. It must be insulated and have heating / cooling capabilities to maintain the proper growth temperature, and be capable of protecting the contents from contamination. Although glass and plastic systems are used for smaller volumes, most large scale units are constructed of metal, generally food quality stainless steel. Stainless steel minimizes corrosion, obviates the adverse effects of metallic ions on cultures, and is accepted by most authorities as suitable for direct contact with food and drugs, for which most of these processes are used. To facilitate cleaning and decontamination, the interior of the tank should be designed to be smooth, without dead legs, ledges, or inaccessible areas. The vessel may need to meet the applicable boiler / pressure vessel requirements, since it may be operated at a slight positive pressure / or be pressurized during a sterilization cycle.

Treatment or filtration of the exhaust from the culture system is not generally warranted for GLSP systems. Beyond that level, the exhaust gases must be filtered or treated. The filters must be capable of removing the organism, allergen, toxin or biologically active compounds present. It may be desirable to pre-treat the exhaust air before filtration by passing it through a treatment device (for example, a condenser, separator, or preheating system), particularly if HEPA rated filters must be used. Note that some guidelines specify the use of HEPA rated filter for all such applications, regardless of the size of the organism being cultured. Where HEPA filters are not specified, the general practice is to use filters to prevent the escape of viable organisms. A single filter is typically used for BL1-LS level control; while two filters in sequence are used for BL2-LS and BL3-LS level control. Most fermentors use an agitator system that is connected to the tank via a rotating seal. Filter housings should have steam on them to maintain a boundary. For higher levels of containment, that is, above BL2-LS, a double mechanical seal should be used. There is some question as to the increased reliability of a double versus a single seal (Hambleton, 1994; Liberman, 1986); however, a double seal is specifically mentioned in the NIH RDNA Guidelines for use where prevention of release is necessary. Where processes involve toxic or biologically active materials, or require additional containment measures because of the agents involved, liquids or steam can be used as the lubricant between the seals. These systems may employ a constant lubricant flow which is sent to a biowaste kill system after exiting the seal. The location of the drive had been a debated issue, that is, bottom vs. top mounted. However, the current design of the bottom-mounted systems provides the necessary containment of the vessel and helps to facilitate maintenance, so are generally used. Magnetically coupled agitation systems may be used for equipment that is small or has minimal mixing needs. Cleanable and aseptic designs are becoming more widely available and are becoming more widely used.

The type of bioreactor used for cell culture depends on whether or not the cells are anchorage dependant (require a surface to attach to in order to survive and propagate). Those that are not anchorage dependent can be grown in vessels very similar to microbial fermentors. They typically use impellers to assure proper mixing of the cells and nutrients, though energy dissipated into the culture via the impellers would be much less than for a microbial culture to avoid the potential for cell damage. Animal cells do not possess the cell wall protection that is typically found on microbes. Other cell culture systems use a bubble column or bubble column with a draft tube, also called an air lift reactor, to achieve proper mixing and aeration of the culture. Using an air perfusion system or a magnetic coupling for the agitator facilitates the containment of the unit; however, their application may be limited by the size of the vessel and / or the viscosity of the media.

If the cells are anchorage dependent, they must be grown in roller bottles, cell factories, microcarriers, hollow fiber systems or other surface attachment-based devices. Most of these systems have integral containment, which improve biosafety, however, although containment of accidental leakage from these systems should be considered based on the volumes and organisms involved.

It is not generally feasible to operate a fermentor or a cell culture vessel under negative pressure due to the obvious problems of foaming and product contamination. For processes where escape from the system must be prevented, the unit should be equipped with devices that monitor the pressure in the chamber and sound an alarm if the set level is exceeded. Depending on the process, foaming can be an issue and the installation of a catch tank or collection system may be warranted to maintain containment.

Pressure vessels must be equipped with a pressure relief system (PRS), which consists of a rupture disc and / or pressure relief valve. Rupture discs are most often used where hygienic design considerations are paramount. When a rupture disc is used for biohazardous organisms, it is desirable to install it in series with a pressure relief valve since the relief is self-sealing after the vessel pressure drops to a safe level. Since the process boundary is breached when the rupture disc opens, the integrity of the disk is monitored. Measures must be in place to safely collect and treat the materials that are released if a rupture disc is ruptured. Devices like a pressure gauge with an alert set-point indicator or an embedded burst indicator in a rupture disc are used to monitor the integrity of the rupture disc. The preventive replacement interval for rupture discs should be set to account for the fatigue that may occur from frequent cycling. A vessel pressure rating (hence rupture disc) significantly higher than process conditions will reduce the likelihood that the relief device will discharge.

From a biosafety standpoint, the criteria above are acceptable for Biosafety Level 1 large scale and GLSP. Additional measures are required for vessels handling large volumes of infectious organisms at a Biosafety Level 2 and above. The NIH RDNA Guidelines require the following:

- The closed system used for the propagation and growth of viable organisms containing the recombinant DNA molecules shall be tested for integrity of the containment features using the organism that will serve as the host for propagating recombinant DNA molecules.

- Testing shall be accomplished prior to the introduction of viable organisms containing recombinant DNA molecules and following modification or replacement of essential containment features.

- Procedures and methods used in the testing shall be appropriate for the equipment design and for recovery and demonstration of the test organism.

- Records of tests and results shall be maintained on file. The closed system used should have sensing devices that monitor the integrity of containment during operations.

- Rotating seals and other mechanical devices directly associated with the closed system used shall be designed to prevent leakage or shall be fully enclosed in ventilated housings that are exhausted through filters which have efficiencies equivalent to high efficiency particulate air (HEPA) filters or through other equivalent treatment devices.

- At BL3-LS, the rotating seals and other mechanical devices directly associated with a closed system shall be designed to prevent leakage or shall be fully enclosed in ventilated housings that are exhausted through filters which have efficiencies equivalent to high efficiency particulate air / HEPA filters or through other equivalent treatment devices.

- In addition, the closed system shall be operated so that the space above the culture level will be maintained at a pressure as low as possible, consistent with equipment design, in order to maintain the integrity of containment features.

Closed aseptic system design is appropriate for primary containment of biological products. Pressure hold testing of the valves and equipment that define the system boundary will ensure system integrity. If dealing with a Risk Group 3 agent, a post-sterilization pressure hold test may also be used. Each system remains closed during processing.

Connections that have been exposed to the culture and that are opened for additions to the process or sample from the process are

- sterilized before connecting the sampling or addition apparatus and
- steamed to inactivate the microorganism before the apparatus is removed.

After processing, systems for Biosafety Level 1-LS and above are decontaminated before opening to the room.

There are many different design approaches employed within the industry to meet the closed-system design criteria. This chapter does not attempt to address them all. However, one example that provides a clear view of the complexities and options available to engineers is sampling systems. Here the need to take in-process samples while maintaining aseptic conditions is of great importance, and, for infectious materials, must be able to collect the sample in a manner which prevents the release of the organisms from the closed system.

One validatable approach to obtaining in-process samples in a closed system design is the implementation of closed disposable sampling as shown in figure 5-12, *Closed Disposable Sampling System*. The NOVAseptic® NovaSeptum system is a pre-sterilized component that has been used to accomplish the same sampling as the previously described process. In this system, components are connected to sample bags via a closed cannula and septum as shown in the following figure. The number of samples that can be taken is limited by the configuration of the bag holder, but the system has proven successful for many sampling functions. The surface of the septum will also be sterilized during the sterilization-in-place (SIP) cycle.

Where biocontainment is required, additional complexity is imposed. The conventional NovaSeal™ mechanical crimping tool does not sterilize the cut tubing when the sample bag is removed. Operator exposure to the culture is not eliminated. Hence an appropriate level of attention to operator training, sample area isolation, and PPE are required. Procedures and appropriate isolation for handling samples are also necessary. Two methods that reduce the concern for personnel exposure to culture at the sampling apparatus are the

- use of a tubing welder to heat seal the tubing and
- use of a re-steamable valve interface for sampling, although this method requires use of equipment that requires cleaning off-line and a conventional sample valve assembly.

When using infectious agents at BSL2-LS or above, the system must be designed to prevent the release of any viable materials from the unit. This may require vessel decontamination before disposable assemblies like the NovaSeptum™ sampling apparatus are removed.

Figure 5-12   Closed Disposable Sampling System

In selecting a sampling approach, there are also a number of variables that will impact the overall cost. These factors include the number of samples required, disposal costs, cleaning frequency, and the level of automation incorporated into the entire process.

5.2.2.2   **Impact on Operations**

It is considered good engineering practice to design the main process stream to be as closed as possible during operation. This will substantially reduce the risk of product contamination from personnel or the environment. It will also provide protection of personnel from potential exposure to viruses or other pathogens that may be present.

There are products and processes that are better suited to closed system design. It is much easier to design closed systems for well-characterized products and well-defined processes that are robust. However, in many cases, such as clinical or process development or smaller scale operations, it may be much easier and cost effective to place the entire process into a classified environment and allow for more open process operations.

Upstream process operations, such as seed inoculum and cell culture unit operations, are typically designed as axenic operations, where the culture is free from living organisms other than the host cell. This, by its nature, implies a closed-system design approach.

5.2.3   Upstream Equipment and Facility Design

5.2.3.1   **Additional Upstream Design Considerations**

Some additional design features are used to ensure the containment of cultures that require secondary containment.

- A second vent filter in series with the primary filter or incinerator may be installed on the vent of large-scale bioreactors and fermentors that are vented outside the processing area.

- Condensate drains from process lines and the drain from vessels that contain active culture are directed to the liquid or effluent biowaste system.

- A separate connection may be provided to the common process drainage system for waste that is not contaminated with the culture. For example, many, if not all, CIP steps typically follow the validated inactivation of the culture so directing CIP waste to the common process drain system substantially reduces the burden on the biological liquid waste inactivation system.

- Inherently redundant designs are selected for components that may leak, like mechanical seals. Double mechanical seals with a pressurized cavity between the seals are typically used for fermentors and bioreactors. With infectious cultures, double mechanical seals may also be considered for pumps.

The balance between containment and GMP is illustrated by seals at equipment joints, like the static O-rings on Ingold style plugs. While installing a back-up O-ring may reduce the likelihood of a leak to the processing room, a leak through the primary O-ring creates a region that may harbor contaminants and will not properly sterilize. Often the choice will be to use a single O-ring to seal the plug. System integrity is then tested in accordance with a GMP procedure that requires pressure testing prior to use. This step plus PPE and secondary containment protect personnel and the environment if a leak occurs.

Bulk and final filling operations are also designed to control contamination and bioburden, both in terms of product contamination and risk. Product contamination at this final stage of the process can have a disastrous impact on patient well being if undetected as well as a tremendous financial impact to the manufacturer.

There are some process unit operations that are difficult to keep fully closed. Some forms of in-process sampling, column fractionation, and solids additions from bulk containers are examples. However, design solutions such as adding bulk powders using an eductor and liquid recirculation loop are practiced. It is also important to understand that closed systems will be open at certain times. This will occur for maintenance and set-up. For example, filter housings are open system components during set up, but are closed during operation. The loss of the closed state due to these routine activities does not negate closure as a key component of the facility design. Therefore, it is necessary to have validated procedures for cleaning, assembly, and sterility testing to ensure that the closed state has been reinstituted. A pressure hold test to prove system integrity will also be used to verify closure of the system prior to operation. A chemical or heat inactivation procedure may be developed to protect personnel before disassembling equipment components like filter housings.

A system design that is truly closed provides a wide array of facility options and a flexible approach to compliance from a regulatory perspective. This is clearly shown in the following quote from ICH Q7A:

> *"Where the equipment itself (for example, closed or contained systems) provides adequate protection of the material, such equipment can be located outdoors."*

The evolution of process system design, leading to current trends in implementation of the Q7A concept of synergy between process and facility design can be seen in a simple look at a case history of facility design.

The traditional approach to clean-room design taken in the manufacture of human therapeutic drugs in the 1980s to early 1990s is shown in the figure 5-13, *Traditional Clean Room Approach*. In this approach, the process space played a critical role in product protection; all process equipment was located inside a controlled environment with the goal of redundancy (also termed a *belts and suspenders* approach) to provide product protection. In this approach, facility cost factors include the sheer size of the room space, equipment and room finishes, HVAC equipment, routine maintenance of clean-room space, cleaning and gowning materials, and increased access control requirements due to open system operations.

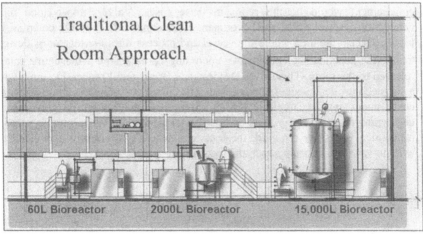

Figure 5-13 Traditional Clean Room Approach

In the 1990s, manufacturing facility cost was increasing at a dramatic rate, as were the costs associated with operation and maintenance of large classified spaces designed to the traditional approach. To offset these costs, the concept of gray space was introduced as shown in the following figure 5-14, *Gray Space Approach*. Here, space segregation between clean and dirty focused on product protection while giving flexibility for maintenance operations by moving many equipment components outside the clean-room envelope.

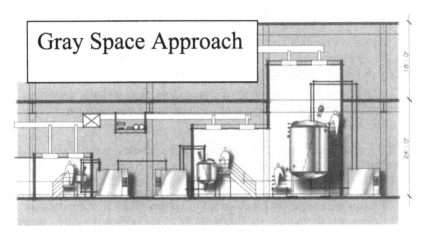

Figure 5-14   Gray Space Approach

By choosing to implement a predominately closed-system approach, the same layout would look dramatically different as seen in the next figure. In this layout, the actual amount of classified space is dramatically reduced. The use of the minimized classified space method cannot be used for large scale cultures of infectious materials that require secondary containment, since secondary containment would be the room in these cases. Valves, non-welded piping connections, vessel ports, and instrument connections all constitute potential leak points. Hence, the area where the vessel and process lines are located is designed to provide secondary containment—including containment if the entire batch is released to the room. This drives the design to apply to all processing units within a single controlled space. Including the entire process system constituting primary containment into a single classified area also may simplify gowning and degowning required for product and personnel protection. When transfers through mechanical areas are required, the transfer lines consist entirely of welded tube with no valves or instruments in the mechanical space.

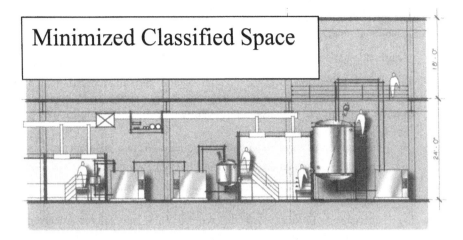

Figure 5-15   Minimizing Classified Space

The final evolution of this case is the true implementation of complete closed-system production. The amount of actual classified clean-room space could be dramatically reduced, and even disappear, which would drastically reduce the costs of operation and maintenance, gowning, and validation for this space. Figure 5-15, *Minimizing Classified Space* provides an example.

5.2.3.2   **Equipment and Facility Integration**

The layout of any biopharmaceutical production facility must be developed around the design of the process unit operations and needs of the facility related to the integration of equipment and process parameters. For new facilities, the ultimate premise is to design the building around the process. By understanding the process, the operations that support the process, the cleanliness and containment requirements, and the operational philosophy, such as gowning and cleaning / changeover, the project team can develop a step-by-step description of the facility from one space to another.

5.2.3.3   **Production Segregation and Flows**

The design of a production facility for biopharmaceutical products must implement the operating philosophy of the plant. The priorities of the design must focus on the regulatory compliance aspect of functionality as it relates to protection of the product. To do this, engineers and architects define functional adjacency related to process unit operations and standard operating procedures.

The protection of the product from external contamination is a key GMP requirement. The concept of segregation is a recognized method to provide this assurance, and it is implemented in the facility design through a variety of avenues. These include physical separation, procedural implementation, environmental means, and chronological scheduling.

Segregation should be viewed from two primary concepts.

1. **Primary segregation** is used to define the overall facility design and organization around environmentally-controlled space envelopes around specific steps in the process (unit operations). Some examples of primary segregation would include

   - physical segregation between the production of two different products,
   - segregation between different lots of the same product,
   - physical segregation between upstream and downstream processing operations, and
   - segregation of process steps that require biological containment from upstream, for example culture media preparation, and downstream processes that follow culture inactivation.

2. **Secondary segregation** focuses on the procedural control of spaces, production activities, and personnel movement. It is normally applied in instances where equipment and components that support the production effort are closed and are protected from the surrounding environment. Some examples of secondary segregation would be

   - segregation of clean and dirty equipment,
   - segregation of personnel flows via airlocks and implementation of gowning activities,
   - chronological separation of activities via scheduling of process activities,
   - segregation of stored materials for quarantine, released, and finished product, and
   - segregation of processing areas to provide separation between processing steps prior to a viral inactivation process step and those steps after the viral inactivation step.

Segregation by spatial means will normally include physically separated areas and dedicated paths of travel for personnel, materials, and equipment. Segregation by time (temporal) would include sequencing the movement of clean and dirty materials and equipment through the same space, but at different times. Segregation by environmental control may include local protection of an open process operation by the use of a classified area.

Open processing operations require segregation, and that is the primary driver behind flow. Flow considerations include the following:

- Direction, type, and quantity / rate of flow
- Cleaning / decontamination
- Transfer requirements

Critical flows that are essential in the evaluation of a sterile production facility are the following:

- Personnel flows: Manufacturing, support, maintenance personnel, etc.
- Material flows: Media and buffer components, raw materials, finished goods
- Equipment flows: Clean / dirty components, portable equipment, product containers
- Product flows: In-process, intermediate, and final
- Waste flow: Solid, liquid, process, and contaminated

The following figure 5-16, *Flow and Segregation Relationships* provides a look at functional adjacency and the relationships of flows and segregation.

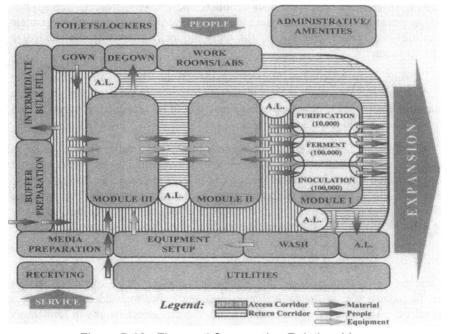

Figure 5-16 Flow and Segregation Relationships

5.2.3.4 **Segregation from a Biosafety Perspective**

Ideally, the facility should be designed so that there is unidirectional flow of materials and personnel. In many cases, that is not possible because it requires a separate way in and out or clean vs. dirty corridors. This is confusing terminology and highlights an area where biosafety and GMP guidance generally diverge.

From a biosafety perspective, dirty signifies the area of the highest concentration of organisms, whereas, in the GMP sense, dirty signifies the crudest form of the product, that is, raw materials. When the material and personnel flow is unidirectional, most of the biosafety and GMP criteria can be met. Facilities that are used to manufacture multiple agents at the same time will need additional features to prevent potential cross contamination. One way to accomplish this is to use an entry-exit corridor system with each of the various rooms or suites having entrance and exit air locks, or to use the double airlock system described previously.

When there is no clean-dirty corridor system, operational procedures need to be used to prevent contamination of the adjacent production areas and potential cross contamination.

Adequate space must be provided for change rooms, storage areas for raw materials, equipment supplies, janitor's closet for housekeeping supplies, equipment cleanup / decontamination, toilets and showers, freezer / refrigerator space, gas supplies / servicing, and other similar purposes. In general, large scale facilities use special gowning, so the entry air lock is typically designed as a change room.

Office areas should be located outside of the large scale facility. It is understood that paperwork areas and computer terminals are necessary in a large scale production area; however, office areas should be separated from production areas by full height walls and doors.

Large scale facilities should be separated from high traffic areas to assist in access restriction and to promote cleanliness. A controlled access system should be considered for all facilities above the GLSP level to protect the product, and at higher levels to protect the personnel from inadvertent exposures. These devices can range from an electronic card entry system, to a combination lock, or a key system. More rigorous systems may be required if the facility security requires it.

5.2.3.5 **Cleaning the Equipment**

Cleaning is a critical issue for equipment design. The equipment design configurations play a dramatic role in cleaning validation and overall process efficiency. The number of penetrations and internal appurtenances in fermentors and bioreactors present an increased cleaning challenge compared to many other unit operations. Current design of clean-in-place systems allows for large-scale fermentation equipment to be effectively cleaned in a manner that can be easily validated for compliance to internationally recognized standards. The figure below shows a typical bioreactor and areas of concern. When initial hot rinses or CIP chemicals are used to inactivate culture, assurance of complete coverage of all hard to clean locations (hard spots) must be validated.

When setting up CIP cycles, consider an ambient water rinse as the first step in the process. Due to the numerous set-up and breakdown requirements for cleaning, an ambient rinse allows confirmation of a closed system without encountering hazardous liquids. Low or no flow alarms on the CIP skid are also recommended.

For equipment that is used for cell culture, where animal derived raw materials may be used, particularly those from bovine sources, the decontamination cycle must be able to achieve 134–138° C for 18 minutes, or as prescribed by the competent authority.

When using organisms at a BSL1-LS or above, if the clean-in-place (CIP) cannot be shown to kill the microorganism grown, all effluents must be piped to a tank for treatment and disposal. Figure 5-17, *Clean-in-Place Hard Spots* shows typical trouble areas.

For some equipment, steaming, treatment with chlorine, other disinfectants, acidic or caustic solution may be used for decontamination. For processes above the GLSP level, the equipment must be decontaminated before opening or cleaning.

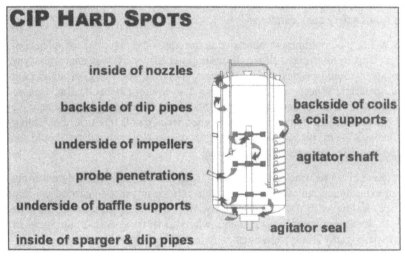

Figure 5-17   Clean-in-Place Hard Spots

5.2.3.6   **Boundary between Biohazardous and Inactivated Culture**

When a biohazardous culture is inactivated before downstream processing, the final unit operation in the biohazard containment area may be an inactivation hold tank. When the culture is chemically inactivated before downstream processing, the inactivating agent may be added and mixed into culture in the bioreactor. The culture is then held for a validated hold period and transferred to the hold tank and held for another validated time required for inactivation. This method ensures exposure of the entire volume of the culture to the inactivating agent.

5.2.4   Downstream Operations

Upstream operations focus on the generation of the product, using cell culture technology as the process platform for manufacturing. Downstream processing focuses on the recovery and purification of the product from the cell culture mixture.

In the downstream processing of the culture material, containment of the material is necessary to protect the product. If the organisms are inactivated in the culture system and there are no toxic, allergenic, or biologically active products (as is the case with GLSP processes), additional containment measures should not be needed. If the cultures are not killed prior to processing, BL1-LS requires that the equipment being used for processing viable organisms be designed to reduce the potential for the escape of viable organisms. At the higher containment levels, the equipment needs to be designed to prevent release. The risk assessment of the organism, which includes an analysis of any harmful characteristics of the organism, is most important consideration in the choice of the containment features of the equipment. BSL-2 and 3 may use agents such as phenol that have their own set of safety considerations.

At BL1-LS, the containment objective is to reduce the potential for release of viable organisms to minimize release. In more practical terms, that means that the equipment used should be designed to prevent spraying, splashing, or significant release of material. Where not designed into the equipment itself, shielding or putting barriers around the equipment or points where this can occur may achieve this level of containment. If the equipment generates aerosols, the use of shielding with an exhaust vent placed near the point of aerosol generation, could be adequate to minimize release to the work area.

The above may be true for work in a research facility when determining whether an experiment can take place on the bench top or must be carried out in a BSC. When dealing with closed systems for BSL-2 LS, the equipment must be designed to ensure there is no leakage whether it is a BSL-2 airborne or bloodborne pathogen. The system must contain both equally.

When dealing with infectious materials at a BSL-2 and above, the goal is to prevent release of aerosols. However, this is where the risk assessment can help determine the extent of the measures to be taken. If the organism involved is transmissible through the blood-borne route, as opposed to being transmissible through respiration, the use of venting, shielding, or barriers may be adequate to prevent employee exposure in the event of equipment malfunction. When dealing with a respiratory transmissible agent, the prevention of aerosol release becomes more critical and more rigorous containment measures must be used.

One way to achieve containment is to place the equipment in a containment device or a room. For example, flow-through centrifuges are known to generate aerosols and where containment is needed, the unit can be placed in a separate room or containment device. Negative pressure isolators may be considered. These can be either flexible or rigid plastic or metal, usually stainless steel. Biological safety cabinets (BSCs) can be used for small equipment that does not generate much turbulence. In some instances, BSC manufacturers, or other specialty equipment fabricators, can make a containment device for specific equipment.

The containment device can consist of plastic shrouding or can be a cabinet with HEPA supply and exhaust to dissipate the heat load. The device needs to provide accessibility to the equipment for operation, routine maintenance and servicing, materials loading and removal, and cleaning. This may necessitate the use of access panels / portholes / gloves. All of these issues need to be thought out in the design of the device so that it can be used in the manner intended.

Through a variety of different unit operations, the volume of material will be reduced by removing unwanted materials from the mixture until the final level of product purity is reached, making the drug ready for final formulation and delivery to the patient. While upstream and downstream operations are very different in their engineering design and operational technologies, they are also inter-related in that there are many aspects of upstream operations that will impact the downstream process. Before discussing the details of downstream operations, it is important to understand what these relational variables are.

- Upstream volume: The volume of material that is delivered from the upstream manufacturing operations has a significant impact on downstream process design. As production volumes have been increasing over the years for protein drugs and antibodies based on rDNA technology, it is not uncommon to see upstream manufacturing bioreactor capacities in the 10,000 to 20,000 L volume. The challenge is how to process such large batch volumes of protein drug solutions through the traditional downstream protein purification steps involving filtration (including various membrane separations for protein retention and buffer exchange) and chromatography unit operations for protein purification. In order to reduce manufacturing costs, volume reduction should be done as soon as possible, and it is likely that this volume reduction will continue to occur as the product is purified.

- Ratio of product to impurities and the nature of the impurities: The more soluble a protein is, the easier it will be to purify. But a large number of proteins are quite insoluble. Changing the pH of the protein's environment can aide in the solubilizing process. Changing the solution's pH can also cause a problem because the solution itself then becomes an impurity that must be removed. The solubilizing process can also cause a protein to become denatured, which further complicates the downstream processing design. The purification process has to enable a high recovery of the target protein if it is to be successful.

5.2.4.1   **Harvest and Recovery**

The recovery and purification of a fermentation or cell culture product is essential to most commercial bioprocess manufacturing processes. The separation of solids such as biomass and insoluble particles from the cell broth is usually the first step in product recovery. The method of product recovery selected will be based on a number of factors; particle size, diffusivity, ionic charge of particles, solubility, surface activity, and density.

There are four major functions in the separation and purification of many enzymes and proteins:

1.   Separation of insoluble products and other solids
2.   Isolation or concentration of the product and removal of water
3.   Purification or removal of contaminating chemicals
4.   Product preparation

The separation of solids from the culture broth is usually the first step. There are four primary methods used for this separation: (1) centrifugation, (2) filtration, (3) coagulation, and (4) flocculation.

Product recovery for protein products retained within the cell (for example, from recombinant *E. coli* cell lines) will require disruption of the cell. Cell disruption can occur from different types of manufacturing operations. The method that is often chosen is a form of mechanical means, where the cell wall is destroyed in order to gain access to the material (protein) of interest. Autolysis, using chemical or biological agents to disrupt or permeablize cell walls or membranes requires consideration of handling and treatment of waste for hazard mitigation. The method chosen must consider the toughness of the cell wall material and the force required to break it down. If the method selected is not performed in a manner that meets the desired expectations, downstream filtration and chromatography operations can become very inefficient due to clogged filters and columns. An additional method of isolating protein from cell, chemical extraction, may use denaturing agents or other hazardous material.

5.2.4.2   **Centrifugation**

Centrifugation is the often the process chosen for this recovery operation. A commercial centrifuge separates the different components of the cell culture mixture by using centrifugal force. A disk-type centrifuge shown in the following figure 5-18, *Disk-type Centrifuge* consists of a large spinning bowl that contains stacked plates. The continuous feed enters the unit from the bottom and the solid waste is discharged from the top of the unit in the clarified liquid. These units are very efficient and are available with clean in place (CIP) and steam in place (SIP) capabilities. However they are also very expensive, and can be very difficult to operate, maintain, and clean. The initial cell lysing and cell separation systems may involve the need to conduct line filter changes associated with the equipment, and there is potential exposure to biological products that will need to be managed through a combination of engineering controls and personal protective equipment.

It is also common for both filtration and chromatography to use NaOH in substantial concentration to sanitize skids, columns, housings, and other equipment.

Figure 5-18   Disk-type Centrifuge

5.2.4.3   **Filtration**

Filtration is probably the most cost-effective method used for the separation of large solids from fermentation broth. The process is based on relative particle size; the molecule of interest may be what passes through the filter, or it may be what is stopped by the filter. It may take multiple filtration steps to successfully remove unwanted particles.

There are a number of types of filtration operations used in modern biotechnology:

- Microfiltration
- Ultrafiltration
- Reverse osmosis

Microfiltration (MF) removes suspended solutes from liquids; generally particles between 0.05 and 5 μm. It involves the retention of particles behind a filter medium while the liquid passes through the filter. Particles are retained because they are larger than the pores in the filter.

Ultrafiltration (UF) separates molecules from solution based on size. The particle size is generally between 0.001 and 0.05 μm. Solute is retained behind the filter while the bulk of the remaining materials pass through the filter. Ultrafiltration membranes have pores small enough to prevent the passage of molecules larger than the molecule of interest. The membranes used also have pores small enough to stop proteins from passing through, but large enough to allow other materials such as salts, buffers, and water to pass. Ultra-filtration systems are generally contained, but there may be in-line filters that require periodic changing and may present opportunity for exposure. Figure 5-19, *Ultrafiltration Skid* provides an image of a typical system.

Figure 5-19   Ultrafiltration Skid (Courtesy of Sartorius)

In reverse osmosis (RO), a pressure is applied to a salt-containing phase, which drives water (solvent) molecules from a low to a high concentration region. This results in the concentration of solute molecules on one side of the filter membrane.

Coagulation and flocculation are usually used to form cell aggregates before centrifugation or filtration to improve the performance of these processes. Coagulation is the formation of small flocs. Many flocculants are multivalent cations such as aluminum, iron, calcium or magnesium. Many polymeric flocculants are used industrially and are available as liquids or solids. These positively charged molecules interact with negatively charged particles and molecules to reduce the barriers to aggregation. In addition, many of these chemicals, under appropriate pH and other conditions such as temperature and salinity, react with water to form insoluble hydroxides which, upon precipitating, link together to form long chains or meshes, physically trapping small particles into the larger floc.

5.2.4.4   **Chromatography**

Chromatography is the separation of different compounds by the flow of a fluid through a column packed with a bed of chromatography resin. Materials move through the bed at different rates of speed based on their chemical properties. As a result of these different migration properties of solutes, components are separated in the form of distinctive bands.

Chromatography can be a very complex operation to perform and control. It is also a very expensive operation due to the fragile nature of the molecules of interest, the cost of the matrix (gel) materials inside the column which can be thousands of dollars per liter of resin, and the highly automated nature of the unit operation. Some of the more important types of chromatographic methods are as follows:

1. Ion-exchange chromatography (IEC): This method is based on the adsorption of ions (or electrically charges compounds) on ion-exchange resin particles by electrostatic forces.

2. Adsorption chromatography (ADC): This method is based on the adsorption of solute molecules onto solid particles, such as alumina and silica gel, by weak van der Waals forces and steric interactions.

3. Affinity chromatography (AFC): This method is based on the specific chemical interactions between solute molecules and ligands bound on support particles. The ligand-solute interaction is very specific. Affinity binding may be molecular size and shape specific.

4. High-pressure liquid chromatography or high performance liquid chromatography (HPLC): This method is primarily for analytical lab use and is based on the general principles of chromatography, with the only difference being high-pressure liquid pressure applied to the packed column. Due to high-pressure liquid, small high surface area particles, and dense column packing, HPLC provides fast and high resolution of solute molecules.

5. Hydrophobic interaction chromatography (HIC): HIC is used to separate proteins on the basis of relative hydrophobicity. Proteins are selectively

adsorbed or eluted from column ligands based on ionic strength of elution buffer applied to column in a gradient.

A chromatography separation process may include more than one of these different methods, some in multiple steps. See figure 5-20, *Chromatography Column Schematic.*

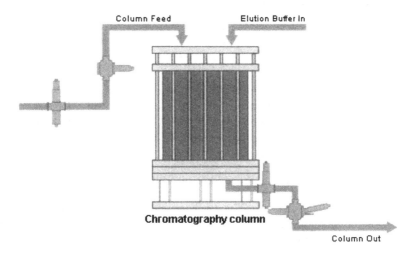

Figure 5-20   Chromatography Column Schematic (Courtesy of Optek)

Chromatographic separations are complex in terms of the science, but the actual steps are relatively simple: column equilibration, sample application, column washing, elution of bound molecules, column regeneration, and re-equilibration. These steps are repeated as long as the column resolution is good and the product is being recovered. As you can expect, column packing materials do have a life cycle that must be monitored to ensure effective product recovery.

Specialty chromatography resins are commonly supplied in a preservative solution of alcohol that prevents the formation of bacteria. Resin is packed into chromatography columns and the columns are frequently washed with alcohol solutions to prevent bacteria growth between campaigns. While it is common to use dilute alcohol solutions (for example, 20% v/v ethanol) to wash columns and elute the resins, engineering designs still have to consider the potential flammability risks involved in purification systems. Volumes of alcohol solutions are generally limited such that standard 55 gallon drums of premixed concentrations are a convenient choice. However, dependent upon plant scale or a particular product demand, some process designs have resulted in the need for bulk alcohol supplies. For these larger demands, concentrated alcohol from a bulk tank can be diluted in a mixing system and delivered to the process use areas by closed piping systems at the desired concentration.

Regardless of the quantity and concentration used, the use of alcohol solutions should be studied by a process hazards analysis evaluation in conjunction with a review of applicable building codes and fire codes. Many such alcohol systems have resulted in requiring high hazard (H-class) occupancy.

5.2.5   Facility Support Issues

All critical equipment and systems supporting the facility should be placed on a preventative maintenance (PM) program. Control panels and items that require regular maintenance should be positioned to allow repair and adjustments to be performed outside of the facility, where possible. Provisions for supply of critical replacement parts should be available, particularly for long lead-time items to prevent extended facility shut down.

HVAC, autoclave, and other equipment and utility support systems should be designed so that maintenance personnel should not have to enter the facility for repairs and scheduled maintenance, especially in facilities where infectious materials are handled.

Sufficient lighting should be provided for all activities, with efforts made to minimize reflections and glare. Lights should be covered with a cleanable surface; lights should be sealed for BL3-LS facilities.

Liquid and gas utility services, if not dedicated to the facility, should be protected with back flow preventers or other devices to prevent contamination (for example, a bump tank for steamable distilled water systems, a liquid disinfectant trap and HEPA or equivalent filter at point of use for vacuum systems).

A separate vacuum system should be considered for any BL2-LS or BL3-LS facilities. If a dedicated system is not available, vacuum lines must be protected with a liquid disinfection trap and appropriate filtration at the point of use.

Provisions should be made to equip the area with telephones, computer terminals, fax machines, and other tools to facilitate information and data transfer outside of the facility and to minimize the need for personnel and paperwork to leave the area. Where required, data packets can be autoclaved with a quick dry vacuum cycle for removal from areas where pathogens are used.

An integrated pest management program should be developed for the facility. Fortunately, the design criteria focusing on cleanability helps in that effort.

All critical systems and equipment should be alarmed. This includes loss of supply air system, loss of exhaust system and containment device exhaust failure. These alarms should be apparent from the outside of the facility so that people will not enter the facility unprepared if the containment has been breached when infectious materials are handled. Critical process equipment must be alarmed when breaches of containment are noted at BL3-LS.

After completion, the facility must be commissioned, that is that there is documented evidence that the facility as built meets the design criteria established. Other terms that have similar meanings include validation, certification or qualification. Regardless of the term used, the items that must be included in the commissioning package include: set of drawings, the defined use and purpose of the facility, the equipment requirements, and test results. The testing must cover: the HVAC system including controls; BSCs, fume hoods and other containment devices; the alarms and failure mode testing; and liquid waste treatment systems and autoclaves. Other controls that may be critical for containment, such as those that monitor pressure on fermentors, must be tested.

5.2.6   Biosafety for Personnel: SOP, Protocols, and PPE

The Bible for biosafety in the US is Biosafety in Microbiological and Biomedical Laboratories (the BMBL), published by the United States Department of Health & Human Services. The book is co-authored by experts from the Centers for Disease Control and Prevention (CDC), the National Institutes of Health (NIH), and recognized individual experts around the country. The book is the most widely recognized set of guidelines on biosafety in laboratories and contains recommendations on biosafety level (BL) classification, risk assessment, select agents, and biological safety cabinets. Outside of the US, there are biosafety guidelines issued by WHO, the UK, the EC, Canada, Belgium, Japan, Australia, and other countries that provide similar guidance.

PPE is the industry acronym for personal protective equipment and refers to items such as lab coats, gloves, and safety glasses. In the pharmaceutical industry and others, this is also referred to as required gowning. Other examples of PPE could include earplugs, coveralls, steel-toed shoes, hard hats and other items in addition to the basic biosafety requirements where industrial or manufacturing process equipment or operations may be encountered.

Personal protective equipment is a part of an overall biosafety strategy, which includes the use of engineering controls and the adoption of aseptic practices, microbiological practices and procedures, as well as, specific biosafety procedures to protect personnel from exposure to biohazardous materials. While the basic PPE is generally recommended for all work at a BL-1 and BL-2, there may be changes required based on the risk assessment of the materials being handled and the processes and equipment involved. Work at biosafety level 3 (BL-3), which could include culturing of HIV, *Mycobacterium tuberculosis* (or other risk group 3 agents that pose serious health risks) may require additional personal protective equipment (for example, back-tie gowns, double gloves, and respirators). Work with risk group 4 agents, such as Ebola virus or other lethal agents, at biosafety level 4 (BL-4), has additional requirements depending on the type of maximum containment facility used. (That is, whether the work will be done Class III biological safety cabinet line or Class II biological safety cabinets used with personnel wearing one-piece positive pressure suits ventilated by a life support system.) Personal protective equipment is the second line of defense to pragmatic, safety-based design.

In addition to PPE requirements, procedures must be established based on the risks of the materials handled. These procedures include standard safe microbiological practices such as washing one's hands after working with biological or chemical materials; no mouth pipetting; safe sharps handling requirements; work surface decontamination; no eating, drinking or smoking inside the laboratory or processing suite; and others. Additional special requirements for waste processing, handling of spills, accidents, and other situations must also be covered in the procedure. Refer to Appendix B, *Large Scale Biosafety Guidelines,* for additional information. Process-specific requirements must be covered in procedures and protocols. Examples include decontaminating disposable products before removal from the work area prior to discarding as waste, cleaning and sterilization of production vessels in between runs of product batches, quarantining of production batches for set periods of time before release of the product to meet quality assurance standards.

6

THE EFFECTS OF EMERGING TECHNOLOGY ON BIOPROCESSING RISK MANAGEMENT

The exploration of biological systems and development of new methodologies, such as genomics and biomolecular design, lead to emerging technologies and opportunities for industrial bioprocessing applications. Production of renewable biofuels such as ethanol and biodiesel, therapeutic stem cells, gene therapy vectors, and new vaccines result in changes in technological applications. Advances in information technology, automation, and process monitoring can enhance product quality and uniformity while increasing cost effectiveness and production. Emerging technologies will provide new, challenging opportunities and risks, as the world shifts from hydrocarbon-based manufacturing to sustainable biomass-based energy sources.

Bioprocessing risk management systems identify risks associated with the emergence of new and existing technologies, providing direction in the management of risks associated with containment or handling of biological agents in the laboratory and manufacturing setting (OSHA Instr. CPL 2-2.45A CH-1 1994). The Center for Chemical Process Safety guideline, *Guidelines for Management of Change for Process Safety*, describes a system for managing such changes in the processing workplace.

6.1   RESEARCHING AND STAYING INFORMED

Bioprocess technology is the basis on which the products of life-science research are translated to a manufacturing environment (Committee on Bioprocess Engineering, 1992). There are emerging technologies in the following very broad areas, some of which could lead to large scale bioprocessing:

- biopharmaceutical,
- renewable-resources, and
- environmental.

6.1.1 Biopharmaceutical

Bioscience research leading to biopharmaceutical manufacturing and production may alleviate many of our most devastating health issues. While the time from research to production is long, and many discoveries never make it, engineers anticipate future needs and design systems to assist in the safe manufacture of the most promising emerging technologies.

6.1.1.1 Drug Discovery and Development

Biopharmaceutical discoveries start in the lab. The discovery and development process of one drug typically costs hundreds of millions of dollars and can span a timeframe of ten to fifteen years or more. The process goal is to ensure that only drugs that are both safe and effective make it to the public. In some rare cases, the Food and Drug Administration (FDA) may choose to fast-track a drug for the treatment of a devastating disease such as the treatment of acquired immune deficiency syndrome.

In the laboratory, scientists identify factors that may cause disease. They look for cellular or genetic determinants of specific illnesses, and then develop biological or chemical drugs that will have a therapeutic effect. Clinical trials test the new drug's efficacy and safety. For every promising new discovery, there are thousands that never make it to market. Examples of other exciting possible advances include the following:

- Treatment of multiple sclerosis through protein optimization technology
- New treatments for bipolar disorders
- New treatments for asthma
- Improved early onset treatments for attention deficit hyperactivity disorder (ADHD)
- Clinical trials for new treatments of Stage III and IV metastatic melanoma
- Drug targets related to the treatment of Alzheimer's disease, cancer treatment therapies, rheumatoid arthritis, and a variety of autoimmune diseases

6.1.1.2 Gene-based Pharmaceuticals

The future is expected to expand on a new generation of drug designs and therapies based on genetic research and the human genome project. Sequencing organisms like the fruit fly will result in more treatments using recombinant DNA technology. This research leads to increasingly effective pharmaceuticals as researchers learn more about the genetic basis for disease and drug response (Collins, Guttmacher, and Guyer, 2003). Dosages and treatments may be customized based on genetic data to alleviate side effects and adverse reactions. The study of genetic factors that influence drug reaction in humans or laboratory organisms is known as pharmacogenetics.

Pharmacogenetics provides unique challenges and unique promise in the clinical research stages of drug development. It blends pharmaceutical sciences such as biochemistry with knowledge of genes, proteins, and single nucleotide polymorphisms. According to *genomics.energy.gov* the possible benefits include the following:

- **"More Powerful Medicines**

  Pharmaceutical companies will be able to create drugs based on the proteins, enzymes, and RNA molecules associated with genes and diseases. This will facilitate drug discovery and allow drug makers to produce a therapy more targeted to specific diseases. This accuracy not only will maximize therapeutic effects but also decrease damage to nearby healthy cells.

- **"Better, Safer Drugs the First Time**

  Instead of the standard trial-and-error method of matching patients with the right drugs, doctors will be able to analyze a patient's genetic profile and prescribe the best available drug therapy from the beginning. Not only will this take the guesswork out of finding the right drug, it will speed recovery time and increase safety as the likelihood of adverse reactions is eliminated. Pharmacogenomics has the potential to dramatically reduce the estimated 100,000 deaths and 2 million hospitalizations that occur each year in the United States as the result of adverse drug response.

- **"More Accurate Methods of Determining Appropriate Drug Dosages**

  Current methods of basing dosages on weight and age will be replaced with dosages based on a person's genetics—how well the body processes the medicine and the time it takes to metabolize it. This will maximize the therapy's value and decrease the likelihood of overdose.

- **"Advanced Screening for Disease**

  Knowing one's genetic code will allow a person to make adequate lifestyle and environmental changes at an early age so as to avoid or lessen the severity of a genetic disease. Likewise, advance knowledge of particular disease susceptibility will allow careful monitoring, and treatments can be introduced at the most appropriate stage to maximize their therapy.

- **"Better Vaccines**

  Vaccines made of genetic material, either DNA or RNA, promise all the benefits of existing vaccines without all the risks. They will activate the immune system but will be unable to cause infections. They will be inexpensive, stable, easy to store, and capable of being engineered to carry several strains of a pathogen at once.

- **"Improvements in the Drug Discovery and Approval Process**

  Biopharmaceutical companies will be able to discover potential therapies more easily using genome targets. Previously failed drug candidates may be revived as they are matched with the niche population they serve. The drug approval process should be facilitated as trials are targeted for specific genetic population groups—providing greater degrees of success. The cost and risk of clinical trials will be reduced by targeting only those persons capable of responding to a drug.

- **"Decrease in the Overall Cost of Health Care**

  Decreases in the number of adverse drug reactions, the number of failed drug trials, the time it takes to get a drug approved, the length of time patients are on medication, the number of medications patients must take to find an effective therapy, the effects of a disease on the body (through early detection), and an increase in the range of possible drug targets will promote a net decrease in the cost of health care" (Human Genome Management Information System (HGMIS) www.ornl.gov/hgmis)[.]

6.1.1.3 **Drug Delivery Research**

Researchers continue to explore drug delivery technology to increase effectiveness and lower risk and patient discomfort.

Plant Virus "Smart Bombs"

At a US university, researchers have modified a common plant virus to transport drugs to only specified cells within the body. The virus would leave the surrounding tissue unaffected. Each one smaller than one thousandth the width of a human hair, these "smart bombs" (North Carolina State University, 2009) may deliver treatment with fewer or no side effects. They could be used to carry chemotherapy to targeted cells.

Researchers indicate that plant virus offers several advantages including ability to survive outside the host, ability to carry chemotherapy to targeted cancer cells, and ability to be easily manufactured.

Bubble Delivery

Internationally, at another university researchers are experimenting with bubbles to deliver chemotherapy. "This development is on the leading edge of the new frontier of drug delivery and cancer treatment," says Prof. Margalit. "Bubble technology can also be applied to other medical conditions, including diabetes, osteoarthritis, wounds, and infectious diseases. In twenty years, it could be widespread" (American Friends of Tel Aviv University, 2009).

6.1.2   Renewable-resources

Can bio-based products replace a sizable portion of petrochemical products? Many large companies believe that to be the case. Rising prices, erratic oil supplies, and increases in energy demands have prompted major oil refiners to research the use of biofuels as alternative energy sources. With research, development, and production, a growing bio-refining industry could take the place traditional fossil-fueled energy.

Agricultural industries are expanding their lines to produce raw materials for sustainable sources of energy. Ethanol innovations have resulted in the commercialization of biofuels and biochemicals made from biosource feedstocks. Research leading to improved enzymes in ethanol production will bring about additional emerging technologies with large scale production of fuel from renewable resources.

The development of renewable, sustainable fuel from agricultural products solves several issues for industrialized countries such as a reduction in pollution, a use for surplus agricultural products, and a useful disposal of food processing wastes.

The U.S. Department of Energy (DOE) and the U. S. Department of Agriculture (USDA) in its report "Biomass as a Feedstock for a Bioenergy and Bioproducts Industry: The Technical Feasibility of a Billion-Ton Annual Supply" indicate that the land resources of the United States are sufficient to produce biomass to displace 30% or more of the nation's present petroleum consumption by the year 2030 (eere.energy.gov / biomass / pdfs / final billionton vision report2.pdf). The report indicates that this is not an upper limit, and that this percentage can be reached without impacting environmentally sensitive areas. A list of biomass products can be found in Chapter 2, *An Overview of the Bioprocessing Industry*.

Biotechnology has had a major impact on agriculture as transgenic crops are expected to produce drugs, chemicals, and fuels. While the biomass resources are available, bioprocess engineering research is needed to address improved saccharification and fermentation of cellulosic solids fraction extracted from a lignocellulosic feedstock so that cost-effective bioproducts can be produced in order to reach DOE and USDA goals.

According to the National Biofuels Action Plan, the DOE has announced plans to invest nearly one billion dollars in conjunction with the private sector and universities to research, develop, and deploy biotechnologies by 2012. Research needs include the following:

- Yield improvement
- Drought resistance
- Harvesting technology
- Pre-processing technology
- Conversion technology
- Reduction in processing costs

There is even research into using the genetically modified physical structure of viruses in nanotechnology energy applications. A Massachusetts Institute of Technology (MIT) team manipulated a gene in the M13 virus to encourage it to coat itself in iron phosphate. Another gene manipulation caused the viruses to bind to carbon nanotubes. This resulted in a highly conductive cathode that may find uses in rechargeable lithium-ion batteries.

6.1.3   Environmental

The environmental aspects of bioprocessing focus primarily in the area of bioremediation. Bioremediation as a discipline is by no means a new concept. Ever since the first organism adapted to take advantage of a human-caused reduction in some environmental characteristic, bioremediation was in effect. However, exploitation of these organisms to purposely remediate environmental damage is relatively recent and new technology to do so is a prime area of research.

6.1.3.1   **Bioprocessing and Waste Management**

Bioremediation uses biological organisms to solve an environmental problem such as contaminated soil or water. While bioprocessing can impact waste management and remediation, additional research is needed to expand bioremediation efforts beyond treatment of wastewater. Imagine the environmental impact of safe biological methods to treat and lessen the volume and toxicity of solid and liquid toxic wastes such as uranium. As an example of an emerging technology, researchers at Oak Ridge National Laboratories are studying the effectiveness of several electron donors for uranium bioremediation in a study funded by the Department of Energy's Environmental Remediation Sciences Program. The resulting report indicates that the particular electron donor chosen affects not only the rate of uranium removal from solution, but also the extent of U6+ conversion to U4+. Results of the study were published in the January–February 2009 issue of the Journal of Environmental Quality (Crop Science Society of America, 2009).

Other possible uses of bioremediation include the following:

- Enzymes to rapidly remove pesticides from irrigation water and soils
- Enzymatic detoxification of nerve gas stockpiles
- Mycelial networks to break down plastics and oil

6.1.4   The Advent of Synthetic Biology

Recently the world's first synthetic self-replicating bacterium was engineered. This breakthrough presents a strong demand for applying risk analysis and assessment skills for any applied technology that arises from newly engineered organisms.

In the hopes that this technology can elicit organisms that produce biofuels, digest toxic spills, clear arteries of cholesterol, produce vaccines in a more timely fashion, grow more efficient crops, and manufacture eco-friendly plastics, we take on the responsibility of assessing all possibilities of risks associated with using this class of organisms. There have been early successes in the field and progress is underway on expanding the applications.

6.2   COMMUNICATING THE IMPACTS OF NEW TECHNOLOGY

Information about new breakthroughs and emerging technologies is communicated by several methods:

- Publication of scientific articles
- Publication of scientific and technical literature
- Employee and employer training and feedback relationships
- Licensing of biotechnical processes
- Plant construction for biorefining or bioprocesses
- Sale of processes or plants
- Education of scientists and engineers by universities and employers
- Technical consulting
- Engineering proposals
- Trade exhibits and conferences
- Publications and meetings of produced by technical organizations
- Governmental studies and publications
- University studies and publications
- Manufacturing protocols and procedures

Technology transfer among government, university, and industry research projects is essential to the discovery and development of emerging technologies. An example of such coordination in the United States, the Bioprocessing Program run by the National Energy Technology Laboratory (NETL), part of the Department of Energy (DOE)'s national laboratory system, supports a mission to advance national, economic, and energy security. According to NETL projects underway or planned include the following:

- Genome sequencing and production of a cost-effective biotoxin for the control of zebra mussels in power plant intakes
- Identification and production of extremophilic / thermophilic microbes to catalyze the conversion of carbon monoxide and water to carbon dioxide and hydrogen for biohydrogen production
- Sampling and characterization of coal utilization by-products (CUBs) from disposal sites for determining the transport and fate of mercury from CUBs to the environment
- Removal, concentration, and recovery of mercury and other heavy metals from coal piles via bacterial action prior to combustion
- Development of more effective sorbents for the removal of mercury from flue gas with the use of mercury-detoxifying thermophiles
- Development of techniques and procedures for industrial use of mercury-free CUBs

Additional information is available at www.netl.doe.gov.

Canadian Biomass Innovation Network (CBIN) coordinates Canada's Federal Government's interdepartmental research, development, and demonstration (RD&D) activities in the area of bioenergy, biofuels, industrial bioproducts and bioprocesses. Additional information is available at www.cbin.ec.gc.ca.

6.2.1   Industry (Communication at Your Site)

Identifying improvements to existing technology or identifying emerging technologies is the job of every employee. Written procedure and protocol systems should consider appropriate methods for communicating information relating to new discoveries and other significant data such as the handling and storage of toxins and biological agents (American Institute of Chemical Engineers, 1996). These systems should include all suitable methods of communication such as meetings, team briefings, signage, training, and reference libraries.

Biotechnology events to share information among academia and industry are held annually. Examples include conferences on topics such as the following:

- Annual Congress of Industrial Biotechnology
  - o   http://www.bio.org/worldcongress/
- Annual World Congress of Industrial Biotechnology

- o   http://www.bit-ibio.com/
- International Congress of Antibodies
  - o   www.bitlifesciences.com/ica2009/
- World Cancer Congress
  - o   www.worldcancercongress.org
- World Summit of Antivirals
  - o   www.bitlifesciences.com/wsa2008
- International Drug Discovery Science and Technology
  - o   http://www.iddst.com/iddst2009/ScientificProgram.asp
- BIT Life Sciences' Annual Congress and Expo of Molecular Diagnostics
  - o   http://www.bitlifesciences.com/cemd2008/
- BioProcess International European Conference & Exhibition
  - o   www.bpi-eu.com/
- The Bioprocessing Summit
  - o   www.healthtech.com/bpd
- Oncology Summit USA
  - o   www.eyeforpharma.com/oncologyusa09
- Next Generation Pharmaceuticals (NGP) Summit
  - o   www.ngpsummit.com/
- Pharmaceutical Manufacturing and Biotechnology Middle East (PABME)
  - o   www.pabme.com/Pabme-Conferences.html
- IBC's BioProcess International Conference & Exhibition (BPI)
  - o   http://www.ibclifesciences.com/bpi/overview.xml
- BIO-IT World Conference & Expo
  - o   www.Bio-ITWorldExpo.com
- The Application of Microspheres, Microparticles & Microbeads in In Vitro Diagnostics and Biotech Applications
  - o   www.genengnews.com/meetings/item.aspx?mid=2441&chid=3
- ACHEMA
  - o   www.achema.de/en/ACHEMA.html
- Molecular Diagnostics Europe
  - o   www.selectbiosciences.com/conferences/MDE
- World Pharmaceutical Congress
  - o   www.worldpharmacongress.com
- Biological Production
  - o   www.biologicalproduction.com
- International Bio Forum & Bio Expo
  - o   www.gate2biotech.com/th-international-bio-forum-bio-expo-japan-2/

- Early to Late Stage Bioprocess Development Summit
  - www.ibclifesciences.com/ETL/overview.xml
- BioProduction
  - www.bio-production.com/
- World Drug Manufacturing
  - www.wdmsummit.com

APPENDIX A – REFERENCES & SELECTED REGULATIONS

Chapter 1

1. Occupational Safety and Health Administration, *Process Safety Management of Highly Hazardous Chemicals, 29 CFR Part 1910, Section 119* (Washington, DC, 1992)

2. Environmental Protection Agency, *Accidental Release Prevention Requirements, Risk Management Programs, Clean Air Act, 40 CFR 68 Section 112 (r)(7)* (Washington, DC, 1996)

3. American Institute of Chemical Engineers, *Guidelines for Implementing Process Safety Management Systems* (Center for Chemical Process Safety, New York, NY, 1994)

4. American Institute of Chemical Engineers, *Guidelines for Process Safety in Outsourced Manufacturing Operations* (Center for Chemical Process Safety, New York, NY, 2000)

5. American Institute of Chemical Engineers, *The Business Case for Process Safety Management* (Center for Chemical Process Safety, New York, NY, 2003)

6. American Chemistry Council, *Resource Guide for the Process Safety Code of Management Practices* (Washington, DC, 1990)

7. World Health Organization, *Laboratory Biosafety Manual*, second ed. (World Health Organization, Geneva, Switzerland, 1993)

8. American Chemistry Council, *Resource Guide for the Process Safety Code of Management Practices* (Washington, DC, 1990)

9. World Health Organization. *Laboratory Biosafety Manual*, second ed. (World Health Organization, Geneva, Switzerland, 1993)

10. *Encyclopedia of Bioprocess Technology*: Fermentation, Biocatalysis, and Bioseparation Volume 1, Michael C. Flickinger, University of Minnesota, St. Paul, Minnesota, Stephen W. Drew, Merck and Co., Inc., Rahway, New Jersey, A Wiley-Interscience Publication (John Wiley & Sons, Inc. New York 1999)

11. Barbara Johnson, PhD, RBP. "Understanding, Assessing, and Communicating Topics Related to Risk in Biomedical Research Facilities, ABSA Anthology of Biosafety IV", *Issues in Public Health*, Chapter 10 (2001)

12. "The Sverdlovsk anthrax outbreak of 1979". M Meselson, J Guillemin, M Hugh-Jones, A Langmuir, I Popova, A Shelokov, and O Yampolskaya, *Science,* 18 November 1994: Vol. 266. no. 5188, pp. 1202–1208

13. CDC/NIH 3rd edition of *Biosafety in Microbiological and Biomedical Laboratories*

14. *New York Times Magazine*, "Symbol Making", November 18, 2001

15. J. A. Perry, Catastrophic Incident Prevention and Proactive Risk Management in the New Biofuels Industry, *AIChE Environmental Progress 7 Sustainable Energy*, Vol. 28, No. 1, pp 72-82, April 2009

Chapter 2

1. *Putting Biotechnology to Work: Bioprocess Engineering*, Committee on Bioprocess Engineering (National Research Council, 1992)

2. *The Bridge, The Role of Bioprocess Engineering in Biotechnology*, Michael Ladisch, Volume 34, Number 3, Fall 2004

3. *Guidelines for Process Safety in Outsourced Manufacturing Operations*, American Institute of Chemical Engineers (Center for Chemical Process Safety, 2000)

4. "Succeed at bioprocess scale-up", John L. Shaw, P.E. and Scott A. Rogers, P.E. (CH2MHill Lockwood Greene, ChemicalProcessing.com, accessed 2008)

5. *Biosafety in Industrial Biotechnology*, edited by P. Hambleton, J. Melling, and T. T. Salisbury (Blackie Academic and professional, 1994)

Chapter 3

1. Occupational Safety and Health Administration, *Process Safety Management of Highly Hazardous Chemicals, 29 CFR Part 1910, Section 119* (Washington, DC, 1992)

2. Deming, W. Edwards, *Out of the Crisis* (MIT Center for Advanced Engineering Study, 1986) ISBN 0-911379-01-0.

3. Environmental Protection Agency, *Accidental Release Prevention Requirements / Risk Management Programs, Clean Air Act, Section 112 (r)(7)* (Washington, DC, 1996)

4. CEN Workshop Agreement, CWA 15793 February 2008 ICS 07.100.01 (English version Laboratory Biorisk Management Standard, February 2008)

5. American Institute of Chemical Engineers, *Guidelines for Implementing Process Safety Management Systems* (Center for Chemical Process Safety, New York, NY, 1994)

6. The Biotechnology Industry Organization (BIO), "Excellence Through Stewardship: Advancing Best Practices in Agricultural Biotechnology" (July 25, 2007)

7. www.epa.gov website, "Wastes - Partnerships - Product Stewardship"

8. Grinsted, J. Risk. "Assessment and contained use of genetically modified organisms" in Tzotzos, G.T. *Genetically modified organisms: a guide to biosafety.* (Wallingford, UK: Centre for Agriculture and Biosciences (CAB) International, 1995;17-35)

9. Cipriano, M.L. "Biosafety considerations for large scale production of microorganisms" in Fleming, D.O., and Hunt, D.L. Biological safety principles and practices. (Washington, D.C.: ASM Press, 2000; 541-55)

10. Collins, C.H. "Safety in microbiology: an overview" in Collins, C.H., and Beale, A.J. *Safety in industrial microbiology and biotechnology* (Oxford, UK: Butterworth-Heinemann Ltd, 1992;1-5)

11. Brunius, N.G.F. in Collins, C.H., and Beale, A.J. *Safety in industrial microbiology and biotechnology* (Oxford, UK: Butterworth-Heinemann Ltd., 1992;239-42)

12. *Laboratory Biosafety Guidelines* 3rd Edition (2004)

13. American Institute of Chemical Engineers, *Guidelines for Investigating Chemical Process Incidents* (Center for Chemical Process Safety, New York, NY, 2003)

14. American Institute of Chemical Engineers, *Guidelines for Process Safety in Outsourced Manufacturing Operations* (Center for Chemical Process Safety, New York, NY, 2000)

15. American Institute of Chemical Engineers, *Guidelines for Risk Based Process Safety*, (Center for Chemical Process Safety, New York, NY, 2007)

15. "OSHA Instruction CPL 2-2.45A CH-1 September 13, 1994 Directorate of Compliance Programs, Subject - 29 CFR 1910.119, Process Safety Management of Highly Hazardous Chemicals" Compliance Guidelines and Enforcement Procedures (OSHA 1994)

16. Johnson, Robert W; Rudy, Steven W; Unwin, Stephen D, *Essential Practices for Managing Chemical Reactivity Hazards, Center for Chemical Process Safety* (Wiley, John & Sons, Incorporated, March 2003)

17. International Organization for Standardization (ISO) for all ISO standards.

Chapter 4

1. Kane, James F. "Environmental assessment of recombinant DNA fermentations" in *Journal of Industrial Microbiology and Biotechnology*, Springer Berlin / Heidelberg, ISSN 1367-5435 (Print) 1476-5535 (Online) (Issue, Volume 11, Number 4 / July, 1993)

2. "Points to Consider in the Characterization of Cell Lines Used to Produce Biologicals." http://www.fda.gov/cber/gdlns/ptccell.pdf Affiliation(s): (2) (Q-One Biotech Ltd., Glasgow, Scotland, UK)

3. *Animal Cell Biotechnology: Methods and Protocols Series: Methods in Biotechnology* | Volume: 8 | Pub. Date: Feb-22-1999 | Page Range: 23-36 | DOI: 10.1385/0-89603-547-6:23

4. Stacey, Glyn N. and Sheeley, Heather J. "Have bio-safety issues in cell culture been overlooked?" in *Journal of Chemical Technology & Biotechnology*, Volume 61, Issue 2 , Pages 95-96 (Society of Chemical Industry 1994)

5. "Animal cell cultures: Risk assessment and biosafety recommendations" http://www.biosafety.be/CU/animalcellcultures/test070205.html

6. "WHO Study Group on Cell Substrates for Production of Biologicals" www.who.int/biologicals/publications/meetings/areas/vaccines/cells/Cells.FIN AL.MtgRep.IK.26_Sep_07.pdf

Chapter 5

1. CDC /NIH Biosafety in Microbiological and Biomedical Laboratories (BMBL)
2. Code of Federal Regulations /Food & Drug Administration: 21 CFR Parts 210 & 211
3. American Institute of Chemical Engineers, *Guidelines for Managing the Security Vulnerabilities of Fixed Chemical Sites* (Center for Chemical Process Safety, New York, NY, 2003)
4. World Health Organization. *Guidelines for the safe production and quality control of IPV manufactured from wild polioviruses.* (Geneva: World Health Organization, 2003)

Chapter 6

1. OSHA Instruction CPL 2-2.45A CH-1 September, 13, 1994 Directorate of Compliance Programs, Subject - 29 CFR 1910.119, Process Safety Management of Highly Hazardous Chemicals - Compliance Guidelines and Enforcement Procedures.
2. American Institute of Chemical Engineers, *Guidelines for Management of Change for Process Safety*, published by The Center for Chemical Process Safety (The American Institute of Chemical Engineers and John Wiley & Sons Inc.; Hoboken, NJ 2008)
3. Committee on Bioprocess Engineering, National Research Council; *Putting Biotechnology to Work: Bioprocess Engineering;* (The National Academies Press, 1992)
4. Collins, Francis S. Green, Eric D. Guttmacher, Alan E. and Guyer., Mark S. "A vision for the future of genomics research" in *Nature* 422, 835-847 (24 April 2003; Received 23 February 2003; Accepted 25 March 2003; Published online 14 April 2003)
5. Human Genome Management Information System (HGMIS) at Oak Ridge National Laboratory for the U.S. Department of Energy Human Genome Program; www.ornl.gov/hgmis
6. North Carolina State University "Plant Virus Nanoparticle Delivers Drug to Cancer Cells" in *Drug Discovery & Development*,.(North Carolina State University, February 13, 2009)
7. American Friends of Tel Aviv University. "New 'Bubble' Targets Only Cancer Cells" (American Friends of Tel Aviv University; Monday, February 9, 2009)
8. See www1.eere.energy.gov/biomass/pdfs/final_billionton_vision_report2.pdf
9. Crop Science Society of America, "Patience Pays Off With Methanol For Uranium Bioremediation" (*Science Daily* 24 February 2009. 17 March 2009) Retrieved from http://www.sciencedaily.com / releases/ 2009 / 02 / 090223121411.htm.

10. American Institute of Chemical Engineers, *Guidelines for Writing Effective Operating and Maintenance Procedures*, (Center for Chemical Process Safety; AIChE, New York, N.Y, 1996)

General References

1. Advisory Committee on Dangerous Pathogens, The large scale contained use of biological agents, 1st Edition, (London, Her Majesty's Stationary Office. 1998)

2. ASM Draft Large Scale Biosafety Guidelines developed by ASM Biosafety Subcommittee under Diane Fleming and Mary Cipriano

3. Cipriano, M. L., Biosafety Considerations for Large Scale Production of Microorganisms, Appendix A, p. 548-554. In D.O. Fleming and D.L. Hunt (ed.), Biological Safety Principles and Practices, 4th ed. (American Society for Microbiology, Washington, D.C.)

4. Cipriano, M.L., "Biosafety Considerations for Design of Large Scale Facilities" p. 179-190 in Richmond, J.Y., (editor), *Anthology of Biosafety* (American Biological Safety Association, Mundelein. 1999)

5. Health and Welfare Canada. *Laboratory Biosafety Guidelines*. Ottawa, Laboratory Centre for Disease Control, 2nd Edition, 1996. http://www.hc-sc.gc.ca/pphb-dgspsp/publicat/lbg-ldmbl-96/index.html

6. Liberman, D.F., Fink, R., and Schaefer, F., "Biosafety in Biotechnology" p.402-408 in Solomon, A.L., Demain, N.A., (eds.), *Industrial Microbiology and* Biotechnology (ASM Press. Washington D.C. 1986)

7. McGarrity,G.J., and Hoerner, C.L., "Biological Safety in the Biotechnology Industry," p. 119-129 in Fleming, D.O., Richardson, J.H., Talies, J.J., and Versley, D. (eds), *Laboratory Safety Principles and Practices. 2nd Edition.* (ASM Press, Washington, DC 1995) Advisory Committee on Dangerous Pathogens, Categorization of Biological Agents According to Hazards and Categories of Containment, 4th Edition, London, Her Majesty's Stationary Office. 1995.

8. Advisory Committee on Dangerous Pathogens, The large scale contained use of biological agents, 1st Edition, (London, Her Majesty's Stationary Office. 1998)

9. Bailey, J.E., and Ollis, D.F., Biochemical Engineering Fundamentals, p.574-634, (McGraw Hill, New York 1977)

10. Centers for Disease Control and Prevention and National Institutes of Health, *Biosafety in Microbiological and Biomedical Laboratories.* 5th Edition. (Washington, D.C. U.S. Department of Health and Human Services. Public Health Service. U.S. Government Printing Office. 2007)

11. Centers for Disease Control and Prevention and National Institutes of Health, *Primary Containment for Biohazardous: Selection, Installation and Use of Biological Safety Cabinets.* 3rd ed. (Washington, D.C. U.S. Department of Health and Human Services. Public Health Services. Centers for Disease Control and Prevention and National Institutes of Health. U.S. Government Printing Office. 1995)

12. Ghidoni, D.A. "HVAC Issues in Secondary Containment" p.63-72 in Richmond, J.Y., (ed.), *Anthology of Biosafety* (American Biological Safety Association, Mundelein. 1999)

13. Hambleton, P., Melling, J., and Salusbury, T.T. (eds), *Biosafety in Industrial Biotechnology*, 1st Edition (Glasgow, Blackie Academic & Professional. 1994)

14. Health and Welfare Canada. *Laboratory Biosafety Guidelines.* (Ottawa, Laboratory Centre for Disease Control, 3rd Edition, 2004)

15. Liberman, D.F., Fink, R., and Schaefer, F., "Biosafety in Biotechnology" p.402-408 in Solomon, A.L., Demain, N.A., (eds.), *Industrial Microbiology and Biotechnology* (ASM Press, Washington DC, 1986)

16. McGarrity,G.J., and Hoerner, C.L., Biological Safety in the Biotechnology Industry, p. 119-129. In Fleming, D.O., Richardson, J.H., Talies, J.J., and Versley, D. (eds), *Laboratory Safety Principles and Practices.* 2nd Edition. (ASM Press, Washington DC, 1995)

17. National Institutes of Health. 2009. Guidelines for Research Involving Recombinant DNA Molecules: http://oba.od.nih.gov/rdna/nih_guidelines_oba.html

18. National Research Council, *Biosafety in the Laboratory.* (Washington D.C., National Academy Press. 1989)

19. Odum, J., "Fundamental Guidelines for Biotech Multiuse Facilities", *Pharmaceutical Engineering.* 1995,15:8-20.

20. Organisation for Economic Co-operation and Development. *Safety Considerations for Biotechnology* (Paris, OECD Publications. 1992)

21. Prime Minister, Guidelines for recombinant DNA experiments. (Tokyo, Ministry of Health, 1991)

22. World Health Organization. *Guidelines for the Safe Production and Quality Control of IPV Manufactured from Wild Polio Virus.* (Geneva: World Health Organization. 2003)

23. World Health Organization. *Laboratory Biosafety Manual.* 3rd Edition. (Geneva: World Health Organization. 2004)

Selected Regulations

Bloodborne Pathogens 29 CFR 1910.1030 (OSHA)

Bloodborne Pathogens Compliance Directive

Hazardous Materials Regulations (49 CFR 100-185)

HazMat Safety

Select Agent Final Rule, Department of Health and Human Services (DHHS)

CDC Correspondence to ABSA regarding Final Rule

Select Agent Final Rule, US Department of Agriculture (USDA)

Hazardous Waste (EPA)

NIH Guidelines on Recombinant DNA Molecules (PDF) April 2002

IBC Resources

Microbiology (EPA)

IAQ (EPA)

Needlestick Standard (OSHA)

Respirator (NIOSH)

Occupational Exposure to Hazardous Chemicals in Laboratories 29 CFR 1910.1450 (OSHA)

Pocket Guide to Chemical Hazards

Personal Protective Equipment (29 CFR 1910.132-139)

Food and Drug Administration (FDA)

Animal Plant Health Inspection Service (APHIS)

Biosafety Compendium on International Regulations & Guidelines

WHO Laboratory Biosecurity Guidance

Material Safety Data Sheets for Infectious Agents (Health Canada)

Draft FDA guidance document on selection of disposable gloves

AIHA Biosafety web link

NIH Division of Safety Publications web site

APPENDIX B – LARGE SCALE BIOSAFETY GUIDELINES

To help ensure appropriate levels of environmental quality and safety for personnel, the following practices should be considered. As the BL level increases, additional requirements or applications and personal protection equipment (PPE) should be incorporated to provide a corresponding increased level of protection for the worker and the product.

This appendix presents the Large Scale Biosafety guideline that was prepared by the former ASM Subcommittee on Laboratory Safety, Task Force on Biosafety that consisted of: M. Cipriano, chair, D. Fleming, R. Hawley, J. Richmond, J. Coggins, B. Fontes, C. Thompson, and S. Wagener. Additional support was received from: P. Meecham, J. Gyuris and R. Rebar, D. Caucheteux, M.E. Kennedy, H. Sheely, S. Gendel, C. Carlson and R. Fink. This Guideline was published in *Biological Safety: Principles and Practices, 4th edition,* by Diane O. Fleming and Debra L. Hunt, available from ASM Press. The chapter on Large Scale Production of Microorganisms, which includes these guidelines, can be found in *Section V. Special Considerations for Biosafety.*

Introduction

Biosafety guidelines for work with small volumes of infectious agents, that is, those amounts typically used for diagnosis, characterization, or basic research, have been established by CDC-NIH and WHO. Additionally, guidelines for working with recombinant DNA molecules in large volumes exist in the NIH Guidelines for Research with Recombinant DNA molecules. However, no specific biosafety guidelines have been established for large scale work with organisms that do not contain recombinant molecules or organism. This document has been developed by including additional equipment and practices for safe large scale work to the existing guidelines from NIH and CDC.

It is understood that the organism, quantity, frequency, and process have a significant impact on the choice of an appropriate biosafety level for the work to be conducted. There is no specific volume which constitutes large scale for microbial agents. Certain CDC-NIH guidance documents have referred to large scale as those volumes typically in excess of those used for identification, typing, assay performance or testing. The risk analysis must include an assessment of the infectivity of the agent, the routes of exposure, the severity of infection, availability of prophylaxis, the level of containment afforded by the process and equipment used, not just the volume of material being handled. Similarly, there is little scientific evidence to support the premise that only volumes greater than 10 liters merit large scale requirements. Certainly that is not true for BL2/BSL-2 and BL3/BSL-3 organisms. The CDC-NIH Guideline recommends raising the biosafety level for culturing and purification of many BL2/BSL-2 organisms, however that was only done in an effort to provide considerations for the Biosafety Officer and scientists in the establishment of the appropriate level of protection.

This document serves as an effort to collect best practices for maximizing the safety for large scale work, and can be used by an Institutional Biosafety Committee and/or a Biological Safety Officer to develop biosafety procedures for the work to be done.

The guidelines will cover 4 different levels for large scale work: Good Large Scale Practices (GLSP), Biosafety Level 1 - Large Scale (BSL1-LS), Biosafety Level 2 - Large Scale (BL2-LS), and Biosafety Level 3 - Large Scale (BL3-LS). The containment conditions for Biosafety Level 4 - Large Scale are not defined here, but should be determined on a case by case basis.

Only the biological hazard of the organism or cell line is addressed here. Other hazards, such as the toxicity or biological activity of the products produced, should be considered separately. These guidelines do not specifically address animal or plant pathogens, however, the containment principles and practices may be useful for some of those agents.

It is recommended that all institutions that engage in large scale research or production with microorganisms appoint a Biological Safety Officer (BSO) to oversee the procedures, facilities and equipment used.

I. Good Large Scale Practices (GLSP)

The GLSP level is recommended for organisms that are not known to cause disease in healthy adults. (Risk Group I), are non-toxigenic, are well characterized and/or have an extended history of safe large scale work. These organisms should not be able to transfer antibiotic resistance to other organisms. Examples of these organisms include *Saccharomyces cerevisiae* CHO cells, *and E. coli* K12. These organisms should have limited survival and/or no adverse consequences if released into the environment.

A. Standard Microbiological Practices

1. Individuals wash their hands after handling viable material.
2. Eating, drinking, smoking, handling contact lenses, and applying cosmetics are not allowed in the work area.

3. Mouth pipetting is prohibited.
4. Work surfaces are capable of being cleaned and disinfected.
5. An insect and rodent control program is in effect.

B. Special Practices

1. Institutions that engage in large scale work should have a health and safety program for their employees.
2. Written instructions and training are provided for personnel who work at GLSP conditions.
3. Processing, sampling, transfer, and handling of viable organisms are done in a manner that minimizes employee exposure and the generation of aerosols.
4. Discharges of viable organisms are disposed of in accordance with applicable local, state, and federal requirements.
5. The facility should have an emergency response plan which includes the handling of spills.

C. Safety Equipment

1. Protective clothing (for example, uniforms and laboratory coats) is provided to minimize the soiling of personal clothing.
2. Safety glasses are worn in the facility. Face shields and/or goggles and face masks are provided for procedures that may involve splashing or spraying of viable organisms.

D. Facilities

1. Sinks, eyewash stations, and safety showers are provided in the work area.

II. Biosafety Level 1 - Large Scale (BL1-LS)

BL1-LS is recommended for the large scale growth of organisms that are not known to cause disease in healthy adult humans and pose minimal hazard to personnel and the environment. These organisms would be handled at BL1/BSL-1 laboratory scale.

A. Standard Microbiological Practices

1. Access to the work area may be restricted at the discretion of the project manager when work is ongoing.

2. Persons wash / clean their hands after they handle viable organisms, after removing gloves, and on leaving the work area.

3. Eating, drinking, smoking, handling contact lenses, and applying cosmetics are not permitted in the work area.

4. Food is stored outside of the work area in cabinets or refrigerators designated and used for this purpose only.

5. Mouth pipetting is prohibited. Only mechanical pipetting devices are used.

6. Work surfaces are decontaminated on a routine basis and after any spill of viable organisms.

7. Procedures are performed carefully in a manner which minimizes aerosol generation.

8. All discharges of the viable organisms are disposed of in accordance with applicable local, state and federal regulations.

9. An insect and rodent control program is in effect.

B. Special Practices

1. Institutions that engage in large scale work have a health and safety program for their employees.

2. Written procedures and training in basic microbiological practices are provided and documented.

3. Medical evaluation, surveillance and treatment are provided where indicated; for example, determine functional status or competency of employees' immune system when working with opportunistic pathogens.

4. Spills and accidents which result in overt exposure to viable organisms are reported to the facility supervisor/manager Medical evaluation, surveillance, and treatment are provided as appropriate and written records are maintained.

5. Emergency plans shall include methods and procedures for handling spills and employee exposures.

6. Cultures of viable organisms are handled in a closed system or other primary containment equipment, for example, biological safety cabinet, which is designed to reduce the potential for the escape of viable organisms.

7. Sample collection and material addition to a closed system, and transfer of culture materials from one closed system to another are conducted in a manner which minimizes employee exposure, the release of viable material and the generation of aerosols.

8. Culture fluids may be removed from a closed system or other primary containment system in a manner which minimizes employee exposure, the release of viable material and the generation of aerosols.

9. Exhaust gases removed from a closed system, or other primary containment system, minimize the release of viable organisms to the environment by the use of appropriate filters or procedures.
10. A closed system or other primary containment equipment that has contained viable organisms shall not be opened for maintenance or other purposes until it has been decontaminated.

C. Safety Equipment

1. Protective clothing (for example, uniforms and laboratory coats) is provided to prevent the contamination or soiling of personal clothing.
2. Safety glasses must be worn. Protective face protection consisting of a face shield, or goggles and face mask is worn for procedures that may involve splashing or spraying of viable organisms.
3. Gloves are worn if the skin on the hands is broken, irritated, or otherwise not intact.

D. Facilities

1. Each work area contains a sink for hand washing, an eyewash station, and an emergency shower. The sink is located near the exit of the work area.
2. The work area has a door which can be closed when large scale work is ongoing.
3. The work area is designed to be easily cleaned.
4. Floors are able to be cleaned and disinfected in case of spills of viable organisms. Rugs are not allowed.
5. Work surfaces are impervious to water and resistant to acids, alkali, organic solvents, and moderate heat.
6. Furniture in the work area is sturdy and placed so that all areas are accessible for cleaning.
7. If the work area has windows that open, they are fitted with fly screens.
8. Facilities are designed to prevent the release of large volumes of viable organisms directly to sewer; for example, floor drains are capped or raised, fitted with liquid tight gaskets to prevent release of untreated organisms to sewer.

III. Large Scale Biosafety Level 2 (BL2-LS)

BL2-LS is recommended for the propagation and cultivation of infectious organisms that would be handled at BL2/BSL-2 in laboratory scale. The following guidelines have been developed for facilities that routinely handle large volumes of these materials.

A. Standard Microbiological Practices

1. Access to the work area is restricted to personnel who meet the entry requirements.
2. Persons wash their hands after they handle viable organisms, after removing gloves and before leaving the work area.
3. Eating, drinking, smoking, handling contact lenses, and applying cosmetics are not permitted in the facility.
4. Food is stored outside of the facility in cabinets or refrigerators designated and used for this purpose only.
5. Mouth pipetting is prohibited. Only mechanical pipetting devices are used.
6. Work surfaces are decontaminated on a routine basis and after any spill of viable organisms.
7. Procedures are performed carefully in a manner which minimizes aerosol generation.
8. All contaminated wastes are decontaminated by an approved method prior to disposal in accordance with local, state, and federal regulations. Wastes that need to be transported to a different area or facility are closed and placed in a durable, leak proof container for transfer. Material to be transferred off site for decontamination is packaged and labeled in accordance with the applicable regulations.
9. All discharges of viable organisms are inactivated by a validated process, that is, one that has been demonstrated to be effective using the organism in question, or with an indicator organism which is known to be more resistant to the physical or chemical methods used; for example, Bacillus stearothermophillus for steam heat.
10. An insect and rodent control program is in effect.

B. Special Practices

1. Institutions that engage in large scale work have a health and safety program for their employees.
2. Doors to the work area are kept closed when work is ongoing.
3. Access to the work area is restricted to personnel whose presence is required and who meet entry requirements, that is, immunization, if any. Individuals who cannot take / do not respond to the vaccine, who cannot take the recommended prophylaxis in the event of an exposure incident, who are at increased risk of infection, or for whom infection may prove unusually hazardous, are not allowed in the work area until their situation has been reviewed by appropriate medical personnel. The individuals are informed of the potential risks and sign an acknowledgement / consent

form, or similar vehicle, which indicates that they understand and accept the potential risk.

4. Written procedures and policies for handling infectious organisms are provided.

5. Personnel are able to demonstrate proficiency in standard microbiological practices and procedures, handling of human pathogens at a Biosafety Level 2. This can consist of previous experience and training. Training in the hazards associated with the organisms involved, and the practices and operations specific to the large scale work area are provided and documented.

6. Appropriate immunizations, medical evaluation surveillance, and treatment are provided where indicated; for example, immunization, and survey of immune status.

7. A hazard warning sign, incorporating the universal biohazard symbol, identifying the infectious agents, listing the name and telephone numbers of the persons knowledgeable about and responsible for the work area, along with any special entry requirements for entering the work area, is posted at the entry to the work area.

8. When appropriate, baseline serum samples or other surveillance samples are collected and stored for all personnel working in or supporting the work area.

9. A biosafety manual is available which details required safety practices and procedures, spill clean-up, handling of accidents, and other appropriate safety information.

10. The use of sharps is avoided. If required, additional safety devices or personal protective equipment are used to prevent accidental exposure. Plasticware is substituted for glassware whenever possible. If glassware is used, it is coated or shielded to minimize the potential for breakage.

11. Viable organisms are placed in a container that prevents leakage during collection, handling, processing, and transport.

12. Viable organisms are handled in a closed system or other primary containment equipment which prevents their release into the environment.

13. Sample collection and material addition to a closed system, and transfer of culture materials from one closed system to another are conducted in a manner which prevents employee exposure and the release of material from the closed system.

14. Culture fluids shall not be removed from a closed system (except as allowed in #13) unless the viable organisms have been inactivated by a validated procedure, or the organism itself is the desired product.

15. Exhaust gases removed from a closed system or other primary containment system are filtered or otherwise treated to prevent the release of viable organisms to the environment.

16. A closed system that has contained viable organisms will not be opened for maintenance or other purposes unless it has been decontaminated.

17. Rotating seals and other mechanical devices directly associated with a closed system used for the propagation of viable organisms are designed to prevent leakage or are fully enclosed in ventilated housings that are exhausted through filters or otherwise treated to prevent the release of viable organisms to the environment

18. Closed systems used for the propagation of viable organisms and other primary containment equipment are tested for the integrity of the containment features prior to use, and following any changes/modifications to the system that could affect the containment characteristics of the equipment. These systems are equipped with a sensing device which monitors the integrity of the containment while in use. Containment equipment for which the integrity cannot be verified or monitored during use, are enclosed in ventilated housings that are exhausted through filters or otherwise treated to prevent the release of viable organisms.

19. Closed systems that are used for propagation of viable organisms or other primary containment equipment are permanently identified. This identifier is used on all records regarding validation, testing, operation, and maintenance.

20. Contaminated equipment and work surfaces are decontaminated with a suitable disinfectant on a routine basis, after spill cleanup. Contaminated equipment is decontaminated prior to servicing or transport. Absorbent toweling / coverings can be used on work surfaces to collect droplets and minimize aerosols and are discarded after use.

21. Individuals seek medical attention immediately after an exposure incident. Spills and accidents that result in overt exposure to infectious materials are immediately reported to the facility supervisor / manager and the BSO. Appropriate medical treatment, medical evaluation, and surveillance are provided, and written records maintained.

22. Emergency procedures include provisions for decontamination and clean-up of all spills/releases of viable material, including proper use of personnel protective equipment.

23. Animals not involved in the work being performed, are not permitted in the work area.

C. Safety Equipment

1. Protective clothing, (for example, lab coats and protective coveralls) is worn to prevent contamination of personal clothing. If the organism can be transmitted through the skin, the protective clothing should be waterproof with a solid-front, for example, wrap-around, or back- or side-tie coats. Protective clothing is removed when leaving the work area.

2. Protective eyewear is worn at all times in the work area. Protective face protection, that is, face shield or goggles and face mask / respirator are worn for any procedures that may involve splashing or spraying. Respirators are worn if the agents involved are respiratory transmissible.

3. Impervious gloves are worn at all times in the work area when work is ongoing. Double gloving and/or the use of latex gloves is considered if personnel are working over extended periods of time, or with processes that may require direct contact with the infectious material. Gloves are discarded upon leaving the work area.

4. The selection of a respirator / face mask is made based on the transmissibility of the agent. If the agent is transmitted through the respiratory route, a respirator with filtration efficiency capable of protecting the individual from the organism is used, for example, HEPA for viruses or N95s for *Mycobacteria tuberculosis*. If the agent is transmitted through mucous membrane contact, a face mask which prevents droplet penetration, for example, plastic molded, is preferred. Personnel are trained in the use of respirators / face masks for procedures that may involve aerosol generation, and for emergency situations that involve the release of viable organisms in the work area.

5. Biological safety cabinets or other ventilated containment devices are used to contain processes of viable organisms if removed from a closed system.

6. Only centrifuge units with sealed rotor heads or safety cups that can be opened in a biological safety cabinet are used; or the centrifuge is placed in a containment device.

D. Facilities

1. Each facility contains a sink for hand washing, an eyewash station, and an emergency shower. The sink is foot, elbow, automatic, or otherwise not hand operated, and located near the door of each room in the work area.

2. The work area has a door which can be closed when large scale work is ongoing.

3. The work area is designed to be easily cleaned and disinfected. Furniture and stationary equipment are sealed to the floor or raised to allow for cleaning and disinfection of the facility.

4. Floors, walls, and ceilings are made of materials that allow for cleaning and disinfection of all surfaces. Light fixtures are covered with a cleanable surface.

5. Work surfaces are impervious to water and resistant to acids, alkali, organic solvents, and moderate heat.

6. Windows to the facility are kept closed and sealed while work is ongoing.

7. General laboratory-type work areas are designed to have a minimum of 6 air changes per hour. For large scale facilities, the number of air changes

per hour will depend on the size of the area, the chemicals and agents handled, the procedures and equipment used, and the microbial and particulate requirements for the area.

8. The ventilation in the work area is designed to maximize the air exchange in the area, for example, the supply and exhaust are placed at opposite ends of the room or using a ceiling supply with lower level exhaust.

9. The work areas in the facility where the infectious organisms are handled are at negative pressure to the surrounding areas.

10. Provisions are made to contain large spills of viable organisms within the facility until appropriately decontaminated. This can be accomplished by placing the equipment in a diked area, or sloping or lowering the floors in those areas to allow for sufficient capacity to contain the viable material and disinfectant.

11. Drainage from the facility is designed to prevent the release of large volumes of viable material directly to sewer, for example, floor drains is capped, raised, or fitted with liquid tight gaskets to prevent release of untreated organisms to sewer.

IV. Large Scale Biosafety Level 3 (BL3-LS)

BL3-LS is recommended for the propagation and cultivation of infectious organisms classified as Risk Group 3.

There is no specific volume which constitutes large scale for infectious agents, since the risks are more dependent on the agent, the procedures being used, and the frequency of activity, as opposed to the volume of the organism being handled. That is why it is recommended that work involving the cultivation or handling of Risk Group 3 organisms, beyond the amounts usually used for the identification or characterization, be done under the direction/auspices of a Biological Safety Officer who can take into account the infectivity of the agent, the routes of exposure, the level of containment afforded by the process and equipment used, the severity of infection, and availability of prophylaxis.

The following guidelines have been developed for facilities that routinely handle large volumes of these materials.

A. Standard Microbiological Practices

1. Access to the facility is restricted to personnel who meet the entry requirements. Individuals, who have not been trained in the operating and emergency procedures of the facility, are accompanied by trained personnel at all times while in the facility.

2. Persons wash their hands after they handle viable materials, after removing gloves and before leaving the work area.

3. Eating, drinking, smoking, handling contact lenses, and applying cosmetics are not permitted in the work area.
4. Food is stored outside of the work area in cabinets or refrigerators designated and used for this purpose only.
5. Mouth pipetting is prohibited. Only mechanical pipetting devices are used.
6. Work surfaces are decontaminated on a routine basis and after any spill of viable material.
7. Procedures are performed carefully in a manner which minimizes aerosol generation.
8. All contaminated wastes are decontaminated by an approved method prior to disposal in accordance with local, state, and federal regulations. Wastes that need to be transported to a different area or facility are closed and placed in a durable, leak proof container for transfer. Material to be transferred off site for decontamination is packaged and labeled in accordance with the applicable regulations.
9. All discharges of the viable materials are inactivated by a validated process, that is, one that has been demonstrated to be effective using the organism in question, or with an indicator organism which is known to be more resistant to the physical or chemical methods used; for example, *Bacillus stearothermophillus* for steam heat.
10. An insect and rodent control program is in effect.

B. Special Practices

1. Institutions that engage in large scale work must have a health and safety program for their employees.
2. Doors to the facility is kept closed except for entry and egress
3. Access to the facility is restricted to personnel whose presence is required and who meet entry requirements, that is, immunization, if any, and comply with all entry and exit procedures. Individuals who cannot take or do not respond to the vaccine, who cannot take the recommended prophylaxis in the event of an exposure incident, who are at increased risk of infection, or for whom infection may prove unusually hazardous, are not allowed in the work area until their situation has been reviewed by appropriate medical personnel. The individuals are informed of the potential risks and sign an acknowledgement / consent form, or similar vehicle, which indicates that they understand and accept the potential risk.
4. Written procedures and policies for handling infectious materials are provided.
5. All personnel working at a BL3-LS facility must demonstrate proficiency in standard microbiological practices and techniques, and in handling human pathogens at a Biosafety Level 3. This can consist of previous

experience and/or training program. Training in the hazards associated with the materials involved, and the specific practices and operations specific to the facility are provided and documented.

6. Appropriate immunizations, medical evaluation surveillance, and treatment are provided where indicated; for example, immunization, and survey of immune status.

7. A hazard warning sign, incorporating the universal biohazard symbol, identifying the infectious agents, listing the name and telephone numbers of the persons knowledgeable about and responsible for the facility, along with any special entry requirements for entering the work area, is posted at the entry to the facility.

8. Baseline serum samples and / or other appropriate specimens are collected and stored for all personnel working in or supporting the facility. Additional specimens may be collected periodically depending on the agents handled

9. A biosafety manual is available which details required safety practices and procedures, spill clean-up, handling of accidents, and other appropriate safety information.

10. The use of sharps is avoided. If required, additional safety devices or personal protective equipment is used to prevent accidental exposure. Plasticware is substituted for glassware whenever possible. If glassware is used, it is coated or shielded to minimize the potential for breakage.

11. Viable organisms are placed in a container that prevents leakage during collection, handling, processing, and transport.

12. Viable organisms are handled in a closed system or other primary containment equipment which prevents their release into the environment.

13. Sample collection and material addition to a closed system, and transfer of culture materials from one closed system to another are conducted in a manner which prevents employee exposure and the release of material from the closed system.

14. Culture fluids are not be removed from a closed system (except as allowed in #13) unless the viable organisms have been inactivated by a validated procedure.

15. Exhaust gases removed from a closed system or other primary containment system are filtered or otherwise treated to prevent the release of viable organisms to the environment.

16. A closed system that has contained viable organisms will not be opened for maintenance or other purposes unless it has been decontaminated.

17. Rotating seals and other mechanical devices directly associated with a closed system used for the propagation of viable organisms are designed to prevent leakage or are fully enclosed in ventilated housings that are exhausted through filters or otherwise treated to prevent the release of viable organisms.

18. Closed systems used for the propagation of viable organisms and other primary containment equipment are tested for the integrity of the containment features prior to use, and following any changes/modifications to the system that could affect the containment characteristics of the equipment. These systems are equipped with a sensing device which monitors the integrity of the containment while in use. Containment equipment for which the integrity cannot be verified or monitored during use, are enclosed in ventilated housings that are exhausted through filters or otherwise treated to prevent the release of viable organisms.

19. Closed systems that are used for propagation of viable organisms or other primary containment equipment are permanently identified. This identifier is used on all records regarding validation, testing, operation, and maintenance.

20. Contaminated equipment and work surfaces are decontaminated with a suitable disinfectant on a routine basis after spill cleanup. Contaminated equipment is decontaminated prior to servicing or transport. Absorbent toweling / coverings can be used on work surfaces to collect droplets and minimize aerosols; however, they should be discarded after use.

21. Individuals seek medical attention immediately after an exposure incident. Spills and accidents that result in overt exposure to infectious materials are immediately reported to the facility supervisor / manager and the BSO. Appropriate medical treatment, medical evaluation, and surveillance are provided, and written records maintained.

22. Emergency procedures include provisions for decontamination and clean-up of all spills/releases of viable material, including proper use of personnel protective equipment.

23. Animals not involved in the work being performed, are not permitted in the work area.

C. Safety Equipment

1. Persons entering the facility will exchange or completely cover their clothing with garments such as solid-front or wrap-around gowns, or coveralls. If the organism can be transmitted through the skin, the protective clothing must be waterproof. Head and shoe covers, or captive shoes are provided. Protective clothing is to be removed when leaving the facility.

2. Protective eyewear is worn at all times in the work area. Protective face protection, that is, face shield or goggles and face mask / respirator are worn for any procedures that may involve splashing or spraying. Respirators are worn if the agents involved are respiratory transmissible.

3. Impervious gloves are worn at all times in the work area when work is ongoing. Double gloving and/or the use of latex gloves is considered if

personnel are working over extended periods of time, or with processes that may require direct contact with the infectious material. Gloves are discarded upon leaving the work area.

4. The selection of a respirator / face mask is made based on the transmissibility of the agent. If the agent is transmitted through the respiratory route, a respirator with filtration efficiency capable of protecting the individual from the organism is used, for example, HEPA for viruses or N95s for *Mycobacteria tuberculosis*. If the agent is transmitted through mucous membrane contact, a face mask which prevents droplet penetration, for example, plastic molded, is preferred. Personnel are trained in the use of respirators / face masks for procedures that may involve aerosol generation, and for emergency situations that involve the release of viable organisms in the work area.

5. Biological safety cabinets Class II or Class III, or other ventilated containment devices are used to contain processes of viable materials if removed from a closed system.

6. Only centrifuge units with sealed rotor heads or safety cups that can be opened in a biological safety cabinet are used; or the centrifuge units are placed in a containment device.

7. Continuous flow centrifuges or other aerosol generating equipment are contained in devices that are exhausted through filters or otherwise treated to prevent the release of viable organisms.

8. Vacuum lines are protected with liquid disinfection traps and HEPA filters or equivalent, which are routinely maintained and replaced as needed.

D. Facilities

1. The facility is separated from areas which are open to unrestricted traffic flow within the building. The entry area to the facility consists of a double doored entry area, such as an airlock or pass-through.

2. Each major work area contains a sink for hand washing, which is not hand operated, for example, automatic, foot, or elbow operated.

3. An eyewash station and emergency shower is available in the facility.

4. The facility is designed to be easily cleaned and disinfected. Furniture and stationary equipment is sealed to the floor, raised, or placed on wheels to allow for cleaning and disinfecting of the facility.

5. Work surfaces are impervious to water and resistant to acids, alkali, organic solvents, and moderate heat.

6. Floors, walls, and ceilings are made of materials that allow for cleaning and disinfection of all surfaces. Light fixtures are sealed, or recessed and covered with a cleanable surface.

7. Penetrations into the containment facility are kept to a minimum and sealed to maintain the integrity of the facility.

8. Windows to the facility are kept closed and sealed.

9. Liquid and gas services to the facility are protected from backflow unless they are dedicated to the facility. Fire protection sprinkler systems do not require backflow preventers.

10. The ventilation system for the facility is designed to control air movement;

   o The position of the supply and exhaust vents is designed to maximize the air exchange in the area, that is, the supply and exhaust are placed at opposite ends of the room or using a ceiling supply with a lower level exhaust.

   o General laboratory-type work areas are designed to have a minimum of 6 air changes per hour. For large scale facilities, the number of air changes per hour will depend on the size of the area, the chemicals and agents being handled, the procedures and equipment used, and the microbial / particulate requirements for the area.

   o The facility is at negative air pressure to the surrounding areas or corridors. The system shall create directional airflow that draws air from the clean areas of the facility into the contaminated areas. If there are multiple contaminated areas, the area of highest potential contamination is the most negative.

   o The exhaust air from the facility is not re-circulated to any other area in the facility, and is discharged to the outside through HEPA filters or other treatments which prevent the release of viable microorganisms.

   o The facility has a dedicated air supply system for the facility. If the supply system is not dedicated to the facility, it contains HEPA filters or appropriate dampers, which can protect the system from potential backflow in the event of a system failure.

   o The supply and exhaust systems for the facility are interlocked to prevent the room pressure from going positive in the event of power or equipment failure. The system is alarmed to indicate system failures or changes in desired air flow.

11. A method for decontaminating all wastes is available in the facility that is, autoclave, chemical disinfection, incineration, or other approved method.

12. Provisions are made to contain large spills of viable organisms within the facility until appropriately decontaminated. This can be accomplished by placing the equipment in a diked area, or sloping or lowering the floors in those areas to allow for sufficient capacity to contain the viable organisms and disinfectant.

13. Drainage from the facility is designed to prevent the release of large volumes of viable organisms directly to sewer, for example, floor drains are capped, raised, or fitted with liquid tight gaskets to prevent release of untreated organisms to sewer.

APPENDIX C – A GENERIC LABORATORY/LARGE SCALE BIOSAFETY CHECKLIST

It is imperative that the local biosafety contact or the biosafety manager is informed and involved in the design phase of the project. The biosafety contact and the project engineer will:

1. Establish a dialog with the customer regarding the nature of biological work to be conducted.
2. Establish the appropriate biosafety level.
3. Choose and specify required/recommended biosafety equipment and various other engineering controls (for example, biological safety cabinets or autoclaves).
4. Discuss and resolve any other biosafety related issues.

The following chart indicates some basic physical requirements for the various biosafety level laboratories and large scale facilities and should serve as a guide only. (Please note: often company biosafety guidelines do not support the construction of BL1/BSL-1 laboratories. The basic biological research laboratory should be constructed to meet or exceed BL2/BSL-2 requirements.)

A Generic Laboratory/Large Scale Biosafety Checklist

Consideration	GLSP	BL1/BSL-1	BL2/BSL-2	BL3/BSL-3	BL4/BSL-4
Isolation of laboratory	No	No	No	Desirable	Yes
Room sealable for decontamination	No	No	No	Yes	Yes
Ventilation:					
Inward air flow	No	No	Yes	Yes	Yes
Mechanical via building system	No	No	Yes	Yes	No
Mechanical, Independent	No	No	No	Desirable	Yes
Filtered air exhaust	No	No	No	Yes	Yes
Double-door entry	No	No	No	Yes	Yes
Airlock	No	No	No	Yes	Yes
Airlock with shower	No	No	No	No	Yes
Effluent treatment	No	No	Yes	Yes	Yes
Autoclave on site	Yes	Yes	Yes	Yes	Yes
Autoclave in laboratory room	No	No	No	Yes	Yes
Double-ended autoclave	No	No	No	Desirable	Yes
Biological Safety Cabinets					
Class I or II	No	No	Yes	Yes	Desirable
Class III	No	No	No	Desirable	Yes

APPENDIX D – BIOLOGICAL ASSESSMENT QUESTIONNAIRE & BIOPROCESS SAFETY CHECKLIST

One company uses forms similar to these one to collect data for performing a biological characterization assessment. Consider your facility's needs and adapt them for your specific bioprocesses.

Biological Assessment Questionnaire
The following questions may provide guidance in assessing and performing a risk assessment for a given project/process:
Section 1. A. - Microorganisms involved:
Does the process involve the use of any microorganisms?   □ YES   □ NO
If NO, proceed to Section 2.
If YES, list the name of the organism and any other relevant information as to its source/history below, (for example, the ATCC number):
What is the Risk Group of the organism?
A. Is the organism infectious?   □ YES   □ NO
If NO, proceed to Section 1. B.

Biological Assessment Questionnaire
What is the standard mode of transmission?
What is the infectious dose, if known?
Is there vaccination or prophylaxis available? ☐ YES ☐ NO
If YES, what is it?
Has the organism been inactivated by a tested procedure during processing? ☐ YES ☐ NO
If YES, provide details on the process and the inactivation data below.
Section 1. B. – RDNA GLSP Aspects:
Has the organism been modified through recombinant DNA technology? ☐ YES ☐ NO
If NO, proceed to Section 2.
Does the organism meet the criteria for GLSP? ☐ YES ☐ NO
What Biosafety Level was assigned to the construct by the institution's RDNA biosafety oversight committee?
Section 2. Cell Lines

Biological Assessment Questionnaire
What types of cells are being used? List cell source species, tissue or cell?
Provide ATCC number or other available information if purchased:
Is it a primary cell line or continuous cell line?
Has the cell line been tested for adventitious agents?
Does it have any viral genetic material integrated into its genome? □ YES □ NO
Section 3. Use of animal materials in the process
Does your project involve animal Sourced materials in the process? □ YES □ NO
If YES, list the type of material , providing the source animal and country of origin:
Section 4. Process related issues
What is the maximum volume of material that you will be processing in a single vessel?
Describe the general process for culture and purification. Indicate whether the entire process will be contained. (attach pages if necessary)
Section 5. Recombinant DNA
Does your project involve any recombinant DNA constructs? □ YES □ NO

Biological Assessment Questionnaire
Do any of the recombinant molecules contain 1/2 or more of the viral genome? □ YES □ NO
If YES, approximately how much?
Does the construct contain genetic material from an infectious agent? □ YES □ NO
If Yes, what is the agent and what portion of the genome is included?
Does the project involve the use of viral vector systems? □ YES □ NO
If Yes, specify the virus and any helper viruses that may be present below:
Does the project involve the transfer of drug resistance to organisms that does not acquire it normally? □ YES □ NO
Does the construct code for any molecules that are toxic? □ YES □ NO
Does the construct code for any materials that are biologically active? □ YES □ NO
Will the construct be used for large scale production purposes? (Note that the volume is generally 10 liters for low-risk work. Work with infectious materials requires review of all cultivation of infectious materials.): □ YES □ NO

Biological Assessment Questionnaire
If Yes, please answer the following to determine if the process qualifies as GLSP:
Is the organism viable? □ YES □ NO
Is the organism non-pathogenic? □ YES □ NO
Is the organism non-toxigenic? □ YES □ NO
Does it involve recombinant strains derived from host organisms that have an extended history of safe large scale use? □ YES □ NO
Does the organism have built-in environmental limitations that permit optimum growth in the large scale setting but limited survival without adverse consequences in the environment? □ YES □ NO
Is the vector insert well characterized? □ YES □ NO
Is the vector insert well characterized? □ YES □ NO
Is the vector insert free from harmful sequences? □ YES □ NO
Is the vector insert limited in size as much as possible to the DNA required to perform the intended function? □ YES □ NO

Biological Assessment Questionnaire
Is the vector insert poorly mobilizable?   □ YES   □ NO
Does the vector insert increase the stability of the construct in the environment?   □ YES   □ NO
Does the vector insert transfer any resistance markers to microorganisms not known to acquire them naturally if such acquisition could compromise the use of a drug to control disease agents in human or veterinary medicine or agriculture?   □ YES   □ NO
Section 6. Biological Toxins
Does your project involve biological toxins?   □ YES   □ NO
List toxin(s) and volumes that will be on site below:
Is this a Select Agent Toxin?   □ YES   □ NO
Additional Comments:

Bioprocess Safety Checklist – Detailed (Project Title and Number) (Product, Process or Method Name)				
PROPERTIES	**DATA (Y/N)**	**DATA GAP CRITICAL (Y/N)**	**COMPLETION DATE FOR OUTSTANDING DATA**	**COMMENTS**
Organism/Cell Line Characteristics				
List name of organism/cell line.				
List designated Biosafety Level of wild type.				
Describe deactivation procedures for organism/cell line.				
Pathogenic to: (circle as applicable) Humans? Animals? Plants?				
Mode of Transmission: (circle as applicable) Inhalation? Ingestion? Injection? Mucus Membrane?				
Is the organism/cell line genetically modified to be a Living Modified Organism (LMO)?				
List designated Biosafety Level of LMO.				
List designated Biosafety level for large scale process of LMO.				
Can LMO cause adverse effects to humans, animals, plants or the environment as a result of establishment or dissemination?				
Does LMO have potential to transfer inserted genetic material				

Bioprocess Safety Checklist – Detailed (Project Title and Number) (Product, Process or Method Name)				
PROPERTIES	**DATA (Y/N)**	**DATA GAP CRITICAL (Y/N)**	**COMPLETION DATE FOR OUTSTANDING DATA**	**COMMENTS**
to other organism?				
Does the LMO have built in biological barriers, which confer limited survivability and replicability, without adverse consequences in the environment?				
General Physical Characteristics of the Product				
List the product.				
Specify physical state of product (for example, protein solution, lyophilized powder, whole broth)				
Specify water solubility.				
Specify pH.				
Toxicological Properties of Product				
Acute toxicity				
Repeat dose toxicity (target organs)				
Skin Irritation				
Eye Irritation				
Skin/respiratory sensitization				
Reproductive effects				
Genetic toxicity				

Bioprocess Safety Checklist – Detailed (Project Title and Number) (Product, Process or Method Name)				
PROPERTIES	DATA (Y/N)	DATA GAP CRITICAL (Y/N)	COMPLETION DATE FOR OUTSTANDING DATA	COMMENTS
Carcinogenicity				
Pharmacological effects				
Occupational/Industrial Hygiene (IH)				
Occupational Exposure Limit				
Occupational Exposure Band				
Industrial hygiene analytical method				
Industrial hygiene sampling method				
Industrial hygiene monitoring data				
Flammability of Liquids in Process				
Flash Point				
Auto-Ignition Temperature				
LOC (Liquids)				
Flammability of Solids In Process				
Train Fire				
Minimum Ignition Temperature - Cloud				
Minimum Ignition Temperature - Layer				
ENVIRONMENT DATA				

Bioprocess Safety Checklist – Detailed (Project Title and Number) (Product, Process or Method Name)				
PROPERTIES	DATA (Y/N)	DATA GAP CRITICAL (Y/N)	COMPLETION DATE FOR OUTSTANDING DATA	COMMENTS
Water solubility				
Dissociation Constants (pKa)				
Vapor Pressure				
Volatility				
Distribution Coefficient (Log Dow)				
Activated Sludge Respiration Inhibition Test (ASRIT)				
Sludge/soil adsorption/desorption				
Microbial toxicity				
Acute toxicity to Daphnia (EC50, 48 hrs)				
Acute toxicity to fish (LC50, 96 hrs)				
Degradation Mechanisms- 1. Photodegradation 2. Biodegradation 3. Hydrolysis				
Occupational Health Program				
Immunization available				
Anti-serum/Anti-toxin				
Special requirements/recommendations (for example pregnancy, immunosuppression)				

APPENDIX E – BIOPROCESS FACILITY AUDIT CHECKLIST

The following questionnaire addresses general biosafety aspects as well as best practices in general for process-related tasks, intermediate steps, or finished products. It can serve both in-house and outsourced facilities. Bioprocessing-related sections or specific items are identified with *(BIO)*. Customize this for your company and the location of the facility.

Generic Bioprocess Facility Audit Checklist			
	YES	NO	NA
Facility Policies Affiliations and Certifications			
Does the company have a documented Responsible Care™ (or equivalent) policy that is fully supported by management?			
Does the company have a documented product stewardship policy, or a health, safety and environmental policy which incorporates the management of chemicals through their total life cycle, thus minimizing adverse effects on human health and well-being and on the environment? If so please attach.			
Is the company a member of a trade or related organization?			
Is any processes' quality management system certified under the requirements of an ISO 9000 or ISO 14000 standard?			
Character of Area around the Plant			
Are there any non-industrial, non-commercial neighbors?			
Is the facility near a major highway, a railroad, waterways, ponds or lakes?			
Are any utilities provided by non-public sources?			
Are there industrial or commercial neighbors that have highly hazardous chemicals, such that a release of one of these chemicals could have an effect on the facility and its people?			
Personnel			

Generic Bioprocess Facility Audit Checklist			
	YES	NO	NA
Is there a documented organizational chart showing departmental responsibilities at the facility? If so, please attach along with a summary of technical, analytical, HS&E professional staff (for example, number of process engineers, project engineers, production engineers).			
Are there training materials for:			
New employees?			
Basic job skills?			
Environmental, Health and Safety			
Statistical Process Control (SPC)?			
Process overviews (when applicable)			
Is training implementation well documented for each employee?			
Is the labor force unionized?			
Are temporary employees used? (Note whether they receive appropriate training.)			
Environmental (BIO)			
Does the facility have written goals in place for pollution prevention or waste minimization?			
Have there been any biologic or chemical spills or releases within the last year?			
Is the groundwater at the facility routinely monitored?			
How is solid waste disposed of:			
Does the facility have pollution control equipment?			
Does the facility have local or country specific permits for			
Air emissions?			
Waste water discharge?			
Waste water treatment, storage and/or disposal?			
Storm water discharge?			
Does the process introduction at this facility create any challenges to regulatory permit limits (air, wastewater or air emissions)?			
Have there been any non-compliance reports sent to a government agency within the last year for:			
Air emissions			

Generic Bioprocess Facility Audit Checklist			
	YES	NO	NA
Waste water discharge?			
Waste treatment, storage and /or disposal?			
Storm water discharge?			
Will the initiation of manufacture or packaging of the materials compromise the facility's sustainability program for reduction of wastes?			
Have any environmental regulatory agencies visited the site during the past 3 years?			
Does an environmental risk management program (such as the USEPA Risk Management Program) apply to this facility?			
Does the facility have environmental catastrophe insurance coverage?			
Does the biological process require special inactivation procedures or infrastructure at this location?			
Maintenance			
Does facility have a documented preventative maintenance program to minimize equipment downtime?			
Is the equipment to be used for manufacturing the products greater than ten years old?			
Does the facility have an inspection program for pressure vessels?			
Does the facility have an inspection program for tanks? (BIO)			
Does the facility have a program for regular inspection and testing of process safety valves and other process safety devices including interlocks?			
Is maintenance provided by in house maintenance or contract maintenance?			
Does the facility have personnel whose duties include reliability engineering responsibilities?			
Security			
Does the facility have the following security features:			
Perimeter security?			
Guard services?			
Controlled entry points?			
Special monitoring features?			
Are employees bound by a confidentiality/ secrecy agreement?			

Generic Bioprocess Facility Audit Checklist			
	YES	NO	NA
Storage			
Is a physical inventory performed on a periodic basis?			
Is electronic data interchange (EDI) available for inventory transactions?			
Are compressed gases stored on site?			
Are any underground storage tanks on site?			
Are all above ground storage tanks within containment systems? (BIO)			
Preparedness, Prevention & Emergency Response			
Have there been any fires or fire department responses at the facility within the last five years?			
Is the responding fire department a paid department? (Note distance.)			
Does the facility have:			
Sprinklered features? (Note if there is containment for the spent sprinkler water.)			
Non-sprinklered features?			
Adequate fire protection water supply?			
Fire pumps			
Fire hydrants?			
Central station supervisory monitor system?			
Pull box alarms?			
Public Address notification warning system?			
Any special fire protection systems?			
On-site fire brigade/emergency response team?			
Hot work permit procedure?			
Are emergency self-contained breathing apparatus (SCBA) or escape air packs available in the work areas involved in the outsourced processing?			
Is employee training on emergency air pack use documented?			
Has the facility conducted an emergency drill within the last two years? (Note date of last drill and whether coordinated with local emergency planning committee.)			

Generic Bioprocess Facility Audit Checklist			
	YES	NO	NA
Is there formal communication between plant management and the local community through a community advisory panel or equivalent? (BIO)			
Does the facility have written operating procedures or GMPs (or equivalent) for emergency shutdowns? (BIO)			
Health & Safety (BIO)			
Does the company have a written health and safety program?			
Does the facility have a biosafety management program and have they identified a biosafety officer accountable for implementation?			
Are any raw materials or finished products handled at the facility considered hazardous?			
Does the facility have a comprehensive incident investigation program that identifies root causes for incidents and tracks corrective actions to completion?			
Has there been reportable lost time injury in the last three years?			
How many recordable injuries have occurred at this facility in each of the last 3 calendar years?			
Has there ever been a fatality at the facility?			
What is the company's experience modification rate with respect to worker's compensation for the last three years?			
Has a safety, health, environmental or regulatory agency visited this facility in the past 5 years?			
Has there been any safety health or environmental violations at the facility within the last five years? (Note date and citation.)			
Are emergency eyewash and shower stations available in the work areas involved in the outsourced processing?			
Does the site have written safety procedures for:			
Use of Respirators?			
Hazard Communication (HAZCOM)?			
Lock out/tag out?			
Confined space entry?			
Line entry?			
Safety meetings? (Note frequency.)			
Natural disasters?			

Generic Bioprocess Facility Audit Checklist			
	YES	NO	NA
Management of Change?			
Is any part of the facility covered under any type of process safety management regulation? (for example, USOHSA Process Safety Management or Control of Major Accident Hazards for EU operations)			
Does the facility conduct documented process hazard analyses on new processes?			
Do plant facilities conform to appropriate area electrical classifications?			
If powdered materials are being handled, does the site evaluate dust explosion issues?			
Are employees required to shower prior to leaving the site for the day?			
Are there provisions for daily uniform changes/cleaning?			
Production			
Is the proposed outsourced activity similar in nature to the facility's core activities? (BIO)			
Does the facility have previous experience in producing the material in question? (Note last production date.)			
Does the facility currently have sufficient idle capacity available for the process and volumes of interest?			
Does the facility currently handle the key raw materials and finished products of interest?			
Is the following material handling equipment used:			
Mass Flow Meters?			
Digital Scales?			
On-line QA Instruments (for example, in-line pH, cell counters, gas chromatographs)			
Is precision of production measuring equipment routinely monitored?			
Is production and packaging equipment routinely calibrated according to a schedule?			
Does the company have an engineering staff which supports the facility?			
Are operating procedures and GMPs reviewed regularly by a technically qualified professional? (BIO)			

Generic Bioprocess Facility Audit Checklist			
	YES	NO	NA
Is there a system to ensure operating procedures and GMPs are current and accurate? (BIO)			
Is there a system to verify that only the current GMPs and operating procedures are used and that obsolete procedures are removed from circulation? (BIO)			
Are all production batches sampled for quality assurance approval? (BIO)			
Are standard production batches with added rework material sampled for quality assurance approval after rework is added?			
Are appropriate directions for reacting to out-of-specification situations documented in standard operating procedures?			
Are detailed records of all problems during production (including non-standard adjustments) kept?			
Are production records kept for the specified retention period for the materials?			
Are specific packaging/repackaging instructions (for example, specific container type, pounds) documented?			
Are detailed records of all problems in packaging kept?			
Are non-conformance summaries sent to facility manager at least quarterly?			
If process is not in statistical control, is there a plan to achieve control?			
Quality Assurance			
Is there a Quality Assurance Policy & Procedure Manual?			
Are SPC/SQC charts maintained and used for continuous product quality improvement?			
Are there documented specifications for all incoming raw materials used in products? (BIO)			
Does the quality assurance group control the approval of raw materials prior to use in production?			
Is incoming material lacking a lot number either rejected or supplied a facility-created lot number?			
Is there a procedure for quarantine of incoming material until approval for use is given by the quality assurance group?			
Are control charts used and monitored for the majority of incoming raw materials?			
Are raw material test results retained for the necessary period?			

Generic Bioprocess Facility Audit Checklist			
	YES	NO	NA
Are raw material suppliers notified of all non-conformances?			
Is evidence of SPC/SQC required of suppliers?			
Are raw material suppliers audited?			
Are analytical test methods documented?			
Is all laboratory equipment calibrated according to a schedule?			
Is precision of all laboratory equipment known?			
Is precision routinely monitored (via control charts or other similar techniques) of all equipment?			
Are repeatability and reproducibility (R&R) studies available for the test procedure?			
Is precision routinely monitored (via control charts or other similar techniques) for the majority of the procedure?			
Are final product specifications on file for all products?			
Are product test results saved at least one year?			
Is there a documented procedure for retaining product samples?			
Is there a documented procedure for quarantine of off-spec materials?			
Does all rework of off-spec materials require documented Quality Assurance authorization?			
Have all production processes (final batch and in-process parameters) been shown to be in statistical control?			
Are there documented specifications for approval to ship final product containers (for example, tank trucks, drums, and bags)?			
Are batch/lot numbers of all shipments recorded?			
Is there a system to confirm that the only material to be shipped is Quality Assurance approved?			
Is there a system to confirm that the material ordered is the right material shipped?			
Are certificates of analyses included with shipping papers?			
Can shipments of a given lot/batch be easily traced to all customers?			
Is there a documented procedure for recalling material from a customer?			
What is the site's experience with cGMPs?			

Generic Bioprocess Facility Audit Checklist			
	YES	NO	NA
What are the results of any governmental (for example, FDA) or quality agency audits during the past three years? (BIO)			
Does the facility have a QC laboratory with analytical capabilities for the tolling contract?			
Does the facility have a process for GMP where applicable? (BIO)			
Potential Exposure			
Are materials handling systems engineered and personal protective equipment used to prevent physical exposure to personnel, including inhalation and skin?			
Is there a certified Industrial Hygienist on staff or accessible within the company or through a consultant?			
Is there a biological safety officer on staff if they are doing large scale production of infectious or recombinant organisms? (BIO)			
Are personnel and air monitoring conducted for chemicals with threshold limit values (TLVs)?			
Is required personal protective equipment easily and readily available and worn by personnel during manufacturing and packaging operations?			
Is spill containment in place around transfer pumps, pipe manifolds, production vessels, packaging lines and storage containers?			
Are there procedures for performing PPE hazard assessments, specifying PPE and training employees on required PPE?			

APPENDIX F – DIRECTIVE 2000/54/EC OF THE EUROPEAN PARLIAMENT AND OF THE COUNCIL

DIRECTIVE 2000/54/EC OF THE EUROPEAN PARLIAMENT AND OF THE COUNCIL of 18 September 2000 on the protection of workers from risks related to exposure to biological agents at work (seventh individual directive within the meaning of Article 16(1) of Directive 89/391/EEC)

ANNEX V I CONTAINMENT FOR INDUSTRIAL PROCESSES

(Article 4(1) and Article 16(2)(a))

Group 1 biological agents For work with group 1 biological agents including life attenuated vaccines, the principles of good occupational safety and hygiene should be observed.

Groups 2, 3 and 4 biological agents It may be appropriate to select and combine containment requirements from different categories below on the basis of a risk assessment related to any particular process or part of a process

A. Containment measures	B. Containment levels		
	2	3	4
1. Viable organisms should be handled in a system which physically separates the process from the environment	YES	YES	YES
2. Exhaust gases from the closed system should be treated so as to:	minimize release	prevent release	prevent release
3. Sample collection, addition of materials to a closed system and transfer of viable organisms to another closed system, should be performed so as to:	minimize release	prevent release	prevent release
4. Bulk culture fluids should not be removed from the closed system unless the viable organisms have been:	Inactivated by a validated means	Inactivated by a validated chemical or physical means	Inactivated by a validated chemical or physical means
5. Seals should be designed so as to	minimize release	prevent release	prevent release

ANNEX V I CONTAINMENT FOR INDUSTRIAL PROCESSES

(Article 4(1) and Article 16(2)(a))

Group 1 biological agents For work with group 1 biological agents including life attenuated vaccines, the principles of good occupational safety and hygiene should be observed.

Groups 2, 3 and 4 biological agents It may be appropriate to select and combine containment requirements from different categories below on the basis of a risk assessment related to any particular process or part of a process

A. Containment measures	B. Containment levels		
	2	3	4
6. Closed systems should be located within a controlled area	OPTIONAL	OPTIONAL	YES, and purpose built
(a) Biohazard signs should be posted	OPTIONAL	YES	YES
(b) Access should be restricted to nominated personnel only	OPTIONAL	YES	YES, via an airlock
(c) Personnel should wear protective clothing	YES, work clothing	YES	a complete change
(d) Decontamination and washing facilities should be provided for personnel	YES	YES	YES
(e) Personnel should shower before leaving the controlled area	NO	OPTIONAL	YES
(f) Effluent from sinks and showers should be collected and inactivated before release	NO	OPTIONAL	YES
(g) The controlled area should be adequately ventilated to minimize air contamination	OPTIONAL	OPTIONAL	YES
(h) The controlled area should be maintained at an air pressure negative to atmosphere	NO	OPTIONAL	YES
(i) Input air and extract air to the controlled area should be HEPA filtered	NO	OPTIONAL	YES
(j) The controlled area should be designed to contain spillage of the entire contents of the closed system	NO	OPTIONAL	YES
(k) The controlled area should be sealable to permit fumigation	NO	OPTIONAL	YES
(l) Effluent treatment before final discharge	Inactivated by a validated means	Inactivated by a validated chemical or physical means	Inactivated by a validated chemical or physical means

ANNEX V I CONTAINMENT FOR INDUSTRIAL PROCESSES

(Article 4(1) and Article 16(2)(a))

Group 1 biological agents For work with group 1 biological agents including life attenuated vaccines, the principles of good occupational safety and hygiene should be observed.

Groups 2, 3 and 4 biological agents It may be appropriate to select and combine containment requirements from different categories below on the basis of a risk assessment related to any particular process or part of a process

A. Containment measures	B. Containment levels		
	2	3	4

ANNEX V I CONTAINMENT FOR INDUSTRIAL PROCESSES

(Article 4(1) and Article 16(2)(a))

Group 1 biological agents For work with group 1 biological agents including life attenuated vaccines, the principles of good occupational safety and hygiene should be observed.

Groups 2, 3 and 4 biological agents It may be appropriate to select and combine containment requirements from different categories below on the basis of a risk assessment related to any particular process or part of a process

A. Containment measures	B. Containment levels		
	2	3	4
1. Viable organisms should be handled in a system which physically separates the process from the environment	YES	YES	YES
2. Exhaust gases from the closed system should be treated so as to:	minimize release	prevent release	prevent release
3. Sample collection, addition of materials to a closed system and transfer of viable organisms to another closed system, should be performed so as to:	minimize release	prevent release	prevent release
4. Bulk culture fluids should not be removed from the closed system unless the viable organisms have been:	Inactivated by a validated means	Inactivated by a validated chemical or physical means	Inactivated by a validated chemical or physical means
5. Seals should be designed so as to	minimize release	prevent release	prevent release
6. Closed systems should be located within a controlled area	OPTIONAL	OPTIONAL	YES, and purpose built
(a) Biohazard signs should be posted	OPTIONAL	YES	YES
(b) Access should be restricted to nominated personnel only	OPTIONAL	YES	YES, via an airlock
(c) Personnel should wear protective clothing	YES, work clothing	YES	a complete change
(d) Decontamination and washing facilities should be provided for personnel	YES	YES	YES
(e) Personnel should shower before leaving the controlled area	NO	OPTIONAL	YES

ANNEX V I CONTAINMENT FOR INDUSTRIAL PROCESSES

(Article 4(1) and Article 16(2)(a))

Group 1 biological agents For work with group 1 biological agents including life attenuated vaccines, the principles of good occupational safety and hygiene should be observed.

Groups 2, 3 and 4 biological agents It may be appropriate to select and combine containment requirements from different categories below on the basis of a risk assessment related to any particular process or part of a process

A. Containment measures	B. Containment levels		
	2	3	4
(f) Effluent from sinks and showers should be collected and inactivated before release	NO	OPTIONAL	YES
(g) The controlled area should be adequately ventilated to minimize air contamination	OPTIONAL	OPTIONAL	YES
(h) The controlled area should be maintained at an air pressure negative to atmosphere	NO	OPTIONAL	YES
(i) Input air and extract air to the controlled area should be HEPA filtered	NO	OPTIONAL	YES
(j) The controlled area should be designed to contain spillage of the entire contents of the closed system	NO	OPTIONAL	YES
(k) The controlled area should be sealable to permit fumigation	NO	OPTIONAL	YES
(l) Effluent treatment before final discharge	Inactivated by a validated means	Inactivated by a validated chemical or physical means	Inactivated by a validated chemical or physical means

APPENDIX G – COMPARISON OF GOOD LARGE SCALE PRACTICE (GLSP) AND BIOSAFETY LEVEL (BL) - LARGE SCALE (LS) PRACTICE

This table is derived from the *NIH GUIDELINES FOR RESEARCH INVOLVING RECOMBINANT DNA MOLECULES (NIH GUIDELINES) Appendix K - Table 1.*

	CRITERION [See Appendix K-VI-B, Footnotes of Appendix K]	GLSP	BL1-LS	BL2-LS	BL3-LS
1.	Formulate and implement institutional codes of practice for safety of personnel and adequate control of hygiene and safety measures.	K-II-A	G-I		
2.	Provide adequate written instructions and training of personnel to keep work place clean and tidy and to keep exposure to biological, chemical or physical agents at a level that does not adversely affect health and safety of employees.	K-II-B	G-I		

CRITERION [See Appendix K-VI-B, Footnotes of Appendix K]		GLSP	BL1-LS	BL2-LS	BL3-LS
3.	Provide changing and hand washing facilities as well as protective clothing, appropriate to the risk, to be worn during work.	K-II-C	G-II-A-1-h	G-II-B-2-f	G-II-C-2-i
4.	Prohibit eating, drinking, smoking, mouth pipetting, and applying cosmetics in the work place.	K-II-C	G-II-A-1-d G-II-A-1-e	G-II-B-1-d G-II-B-1-e	G-II-C-1-c G-II-C-1-d
5.	Internal accident reporting.	K-II-G	K-III-A	K-IV-A	K-V-A
6.	Medical surveillance.	NR	NR		
7.	Viable organisms should be handled in a system that physically separates the process from the external environment (closed system or other primary containment).	NR	K-III-B	K-IV-B	K-V-B
8.	Culture fluids not removed from a system until organisms are inactivated.	NR	K-III-C	K-IV-C	K-V-C
9.	Inactivation of waste solutions and materials with respect to their biohazard potential.	K-II-E	K-III-C	K-IV-C	K-V-C
10.	Control of aerosols by engineering or procedural controls to prevent or minimize release of organisms during sampling from a system, addition of materials to a system, transfer of cultivated cells, and removal of material, products, and effluent from a system.	Minimize Procedure K-II-F	Minimize Engineer K-III-B K-III-D	Prevent Engineer K-IV-B K-IV-D	Prevent Engineer K-V-B K-V-D

CRITERION [See Appendix K-VI-B, Footnotes of Appendix K]		GLSP	BL1-LS	BL2-LS	BL3-LS
11.	Treatment of exhaust gases from a closed system to minimize or prevent release of viable organisms.	NR	Minimize K-III-E	Prevent K-IV-E	Prevent K-V-E
12.	Closed system that has contained viable organisms not to be opened until sterilized by a validated procedure.	NR	K-III-F	K-IV-F	K-V-F
13.	Closed system to be maintained at as a low pressure as possible to maintain integrity of containment features.	NR	NR	NR	K-V-G
14.	Rotating seals and other penetrations into closed system designed to prevent or minimize leakage.	NR	NR	Prevent K-IV-G	Prevent K-V-H
15.	Closed system shall incorporate monitoring or sensing devices to monitor the integrity of containment.	NR	NR	K-IV-H	K-V-I
16.	Validated integrity testing of closed containment system.	NR	NR	K-IV-I	K-V-J
17.	Closed system to be permanently identified for record keeping purposes.	NR	NR	K-IV-J	K-V-K
18.	Universal biosafety sign to be posted on each closed system.	NR	NR	K-IV-K	K-V-L
19.	Emergency plans required for handling large losses of cultures.	K-II-G	K-III-G	K-IV-L	K-V-M

CRITERION [See Appendix K-VI-B, Footnotes of Appendix K]	GLSP	BL1-LS	BL2-LS	BL3-LS
20.	Access to the work place.	NR	G-II-A-1-a	G-II-B-1-a	K-V-N
21.	Requirements for controlled access area.	NR	NR	NR	K-V-N&O

NR = not required

Appendix K-VI. - Footnotes of Appendix K

Appendix K-VI-A. This table is derived from the text in Appendices G (Physical Containment) and K and is not to be used in lieu of Appendices G and K.

Appendix K-VI-B. The criteria in this grid address only the biological hazards associated with organisms containing recombinant DNA. Other hazards accompanying the large scale cultivation of such organisms (for example, toxic properties of products; physical, mechanical, and chemical aspects of downstream processing) are not addressed and shall be considered separately, albeit in conjunction with this grid.

Appendix K-VII.   Definitions to Accompany Containment Grid and Appendix K

Appendix K-VII-A. Accidental Release. An accidental release is the unintentional discharge of a microbiological agent (that is, microorganism or virus) or eukaryotic cell due to a failure in the containment system.

Appendix K-VII-B. Biological Barrier. A biological barrier is an impediment (naturally occurring or introduced) to the infectivity and/or survival of a microbiological agent or eukaryotic cell once it has been released into the environment.

Appendix K-VII-C. Closed System. A closed system is one in which by its design and proper operation, prevents release of a microbiological agent or eukaryotic cell contained therein.

Appendix K-VII-D. Containment. Containment is the confinement of a microbiological agent or eukaryotic cell that is being cultured, stored, manipulated, transported, or destroyed in order to prevent or limit its contact with people and/or the environment. Methods used to achieve this include: physical and biological barriers and inactivation using physical or chemical means.

Appendix K-VII-E. De minimis Release. De minimis release is the release of: (i) viable microbiological agents or eukaryotic cells that does not result in the establishment of disease in healthy people, plants, or animals; or (ii) in uncontrolled proliferation of any microbiological agents or eukaryotic cells.

Appendix K-VII-F. Disinfection. Disinfection is a process by which viable microbiological agents or eukaryotic cells are reduced to a level unlikely to produce disease in healthy people, plants, or animals.

Appendix K-VII-G. Good Large Scale Practice Organism. For an organism to qualify for Good Large Scale Practice consideration, it must meet the following criteria [Reference: Organization for Economic Cooperation and Development, Recombinant DNA Safety Considerations, 1987, p. 34-35]: (i) the host organism should be non-pathogenic, should not contain adventitious agents and should have an extended history of safe large scale use or have built-in environmental limitations that permit optimum growth in the large scale setting but limited survival without adverse consequences in the environment; (ii) the recombinant DNA-engineered organism should be non-pathogenic, should be as safe in the large scale setting as the host organism, and without adverse consequences in the environment; and

(iii) the vector/insert should be well characterized and free from known harmful sequences; should be limited in size as much as possible to the DNA required to perform the intended function; should not increase the stability of the construct in the environment unless that is a requirement of the intended function; should be poorly mobilizable; and should not transfer any resistance markers to microorganisms unknown to acquire them naturally if such acquisition could compromise the use of a drug to control disease agents in human or veterinary medicine or agriculture.

Appendix K-VII-H. Inactivation. Inactivation is any process that destroys the ability of a specific microbiological agent or eukaryotic cell to self-replicate.

Appendix K-VII-I. Incidental Release. An incidental release is the discharge of a microbiological agent or eukaryotic cell from a containment system that is expected when the system is appropriately designed and properly operated and maintained.

Appendix K-VII-J. Minimization. Minimization is the design and operation of containment systems in order that any incidental release is a de minimis release.

Appendix K-VII-K. Pathogen. A pathogen is any microbiological agent or eukaryotic cell containing sufficient genetic information, which upon expression of such information, is capable of producing disease in healthy people, plants, or animals.

Appendix K-VII-L. Physical Barrier. A physical barrier is considered any equipment, facilities, or devices (for example, fermentors, factories, filters, thermal oxidizers) which are designed to achieve containment.

Appendix K-VII-M. Release. Release is the discharge of a microbiological agent or eukaryotic cell from a containment system. Discharges can be incidental or accidental. Incidental releases are de minimis in nature; accidental releases may be de minimis in nature.

GLOSSARY

A

Adventitious agents — viruses and toxins, often infectious agents, that can accidentally contaminate a cell line.

Adverse reaction — undesirable effect of a drug, vaccine, or medical device; it can be as mild as a short-term injection-site irritation or as serious as a life-threatening acute onset of anaphylaxis; also referred to as adverse event.

Agent — a microorganism or chemical substance, the presence or absence of which triggers a particular disease or infection.

Anaphylactic shock, anaphylaxis — an immediate, severe, potentially life-threatening 'shock' reaction to an allergen (drug substance, food, or insect sting) resulting in breathing difficulty, a drop in blood pressure, and/or unconsciousness.

Antibiotic — a drug that fights a bacterial infection.

Antibodies (abs) — infection-fighting proteins that the body produces to destroy foreign microorganisms or toxins (antigens); also known as immunoglobulins.

Antigen (also immunogen) — a foreign substance in the body (a bacterium, virus, or protein, for example) that can cause disease and whose presence triggers an immune response (the formation of antibodies).

Antigenicity — the relative ability of a substance to function as an antigen.

Antimicrobial agents — substances such as antibacterial, antiviral, antifungal, and antiparasitic drugs that kill disease-causing organisms.

Attenuated vaccines — a disease-causing virus is weakened (attenuated) by chemicals, aging, or nutrient deprivation to trigger a strong immune response without being able to cause the disease itself.

B

Biological oxygen demand — the amount of dissolved oxygen in water, given in lbs. (kgs) or % that is consumed by biological oxidation of a chemical

Bioprocessing — Bioprocessing makes use of microorganisms, cells in culture or enzymes to manufacture products.

Bioprocess engineering — the sub-discipline within biotechnology that is responsible for translating the discoveries of life science into practical products, processes, or systems that can serve the needs of society.

Biosafety level — A specific combination of work practices, safety equipment, and facilities designed to minimize the exposure of workers and the environment to infectious agents.

Broth (fermentation/cell culture) — the cells, nutrients, waste, and other components that make up the contents of a microbial bioreactor.

C

Coccus (plural cocci) — a spherical bacterium.

D

DNA vaccine (nucleic acid vaccine) — injection of a gene that codes for a specific antigen, enabling the recipient to produce that antigen directly to achieve the desired immune response.

E

Edible vaccines — plants or vegetables (such as potato) that are engineered to express an antigenic protein. Upon consumption, the protein is recognized by the immune system and vaccinates the patient against the original antigen.

Efficacy — proven ability of a drug or vaccine to produce a desired clinical effect at the optimal dose.

Endemic — the continual presence of a disease in a population.

Epidemic — occurrence of a disease within a specific area or region in excess of its normal level.

Epitope — a molecular region on the surface of an antigen capable of eliciting an immune response and of combining with the specific antibody produced by such a response called also *determinant, antigenic determinant.*

Expression system — in genetic engineering, the cells (host organism) into which a gene is inserted to manufacture desired proteins; the gene is combined with a genetic vector (such as a virus or circular DNA molecule called a plasmid) to provide the genetic context in which it will function in the cell; that is, the gene will be expressed as the protein of interest.

F

Fungus (plural fungi) — occurring as single-celled forms such as yeast and complex forms such as mushrooms, fungi can cause infections from mild skin infections such as ringworm and athlete's foot to life-threatening conditions such as cryptococcal meningitis, histoplasmosis, and clastomycosis.

G

Genome — an organism's complete set of genetic material.

H

Hazardous energy control procedure — written procedure for affixing appropriate lockout devices or tag out devices to energy isolating devices, and to otherwise disable machines or equipment to prevent unexpected energization, start up or release of stored energy in order to prevent injury to employees.

Host cell — a cell whose metabolism is used by a virus for growth and reproduction or into which a plasmid is introduced in recombinant DNA experiments; in bioprocessing, the cells engineered and cultured to express a protein of interest are the expression system host cells.

Host factor — the intrinsic factor such as age, race, sex, or behavior that influences a person's exposure, susceptibility, or response to a causative agent.

HVAC — the heating, ventilation, and air conditioning system of a building.

I

Immune response — the reaction of the immune system to foreign organisms.

Immunity — measured by the presence of antibodies in the blood, a natural or acquired resistance to a specific disease, whether partial or complete, specific or nonspecific, lasting or temporary.

Immunodeficiency (also immunosuppression) — condition in which the normal immune response is weakened or diminished from the effects of disease or drugs.

Infectious agents — organisms (such as bacteria, viruses, and fungi) that cause disease.

Inoculation — introducing material (a vaccine, for example) into the body's tissues; also introducing cells into a culture medium.

In vitro — (literally, within glass) an activity taking place in a controlled environment outside of a living organism

In vivo — (literally, within the living) an activity taking place within a living organism

J

Joint committee on vaccination and immunization (JCVI) — non-departmental public body and statutory expert standing advisory committee that advises the secretaries of state for health in Scotland, Wales, and Northern Ireland on matters relating to communicable diseases that are potentially preventable through immunization.

L

Live-vector vaccine — a vaccine that stimulates an immune response by using a non-disease-causing bacterium or virus to transport foreign genes into the body.

M

Management of change (MOC) a management system for ensuring that changes to processes are properly analyzed (for example, for potential adverse impacts), documented, and communicated to affected personnel.

Mechanical integrity (MI) a management system for ensuring the ongoing durability and functionality of equipment.

Monoclonal antibody (MAb) — a highly specific, purified antibody that recognizes only a single epitope.

N

Nutraceutical — extracts of foods claimed to have a medicinal effect on human health.

O

Oncogene — an oncogene is a gene that, when mutated or expressed at high levels, helps turn a normal cell into a tumor cell.

Outbreak — the spread of a disease over a short period in a limited geographic area.

P

Parenteral — a medicine that is administered intravenously or by injection; most vaccines can be administered subcutaneously (to the fatty layer immediately below the skin) or intramuscularly — but not intravenously.

Performance measure — a metric used to monitor or evaluate the operation of a program activity or management system.

Personal protective equipment (PPE) — equipment designed to protect employees from serious workplace injuries or illnesses resulting from contact with chemical, radiological, physical, electrical, mechanical, or other workplace hazards. Besides face shields, safety glasses, hard hats, and safety shoes, PPE includes a variety of devices and garments such as goggles, coveralls, gloves, vests, earplugs, and respirators.

Pharmacogenomics — a biotechnological science that combines the techniques of medicine, pharmacology, and genomics and is concerned with developing drug therapies to compensate for genetic differences in patients which cause varied responses to a single therapeutic regimen.

Pre-startup safety review (PSSR) — a final check, initiated by a trigger event, prior to the use or reuse of a new or changed aspect of a process. It is also the term for the OSHA PSM and EPA RMP element that defines a management system for ensuring that new or modified processes are ready for startup.

Process hazard analysis (PHA) a systematic evaluation of process hazards with the purpose of ensuring that sufficient safeguards are in place to manage the inherent risks.

Process safety information (PSI) a compilation of chemical hazard, technology, and equipment documentation needed to manage process safety.

Prophylaxis — prevention of a disease or condition.

Proteomics — the study of protein functions and structure by organisms.

Protocol — documentation that ties together all SOPs (standard operating procedures) to direct the work performed in a regulated facility —who approves what, who is allowed to sign off on materials and products, where certain files and documents are kept, and so on; also a detailed plan for a clinical trial, stating (among other things) the trial's rationale, purpose, scope, dosages, routes of administration, length of study, and eligibility criteria.

Protozoa — single-celled organisms, many of which are parasitic (for example, the malaria parasite).

Q

Quality assurance (QA) activities to ensure that equipment is designed appropriately and to ensure that the design intent is not compromised throughout the equipment's entire life cycle.

R

Replacement in kind — replacement that satisfies the design specifications.

Risk — measure of potential loss (for example, human injury, environmental insult, economic penalty) in terms of the magnitude of the loss and the likelihood that the loss will occur.

Risk analysis — the development of a qualitative and/or quantitative estimate of risk-based on engineering evaluation and mathematical techniques (quantitative only) for combining estimates of event consequences, frequencies, and detectability.

Risk, risk factor — along with the probability that an event will occur (risk) are those factors of behavior, lifestyle, environment, or heredity associated with increasing or decreasing that probability.

S

Scale-up — the steps involved in transferring a manufacturing process or section of a process from laboratory scale to the level of commercial production.

Select agents — organisms that are of particular concern to the federal government because of their potential use as biological weapons.

Sparging — spraying; a sparger on a fermentor sprays air into the broth.

Strain — a genetic variant within a species.

Substrate (in cell culture) — surface on which a cell or organism grows or is attached—such as the use of microcarriers in cell culture; most eukaryotic cell types require attachment to a substrate for survival; also called extracellular matrices; in the body they are composed mainly of proteins and provide chemical cues that affect or guide the behavior of cells.

T

Trigger event — any change being made to an existing process, or any new facility being added to a process or facility, or any other activity a facility designates as needing a pre-startup safety review. One example of a non-change related trigger event is performing a PSSR before restart after an emergency shutdown.

V

Vector — a plasmid, virus, or other vehicle for carrying a DNA sequence into the cells of another species; also a method (such as genetically engineered viruses or bacteria) of delivering genetic material to cells.

Verification activity — a test, field observation, or other activity used to ensure that personnel have acquired necessary skills and knowledge following training.

Virus — a microorganism that grows and reproduces in living cells of a host (bacteria, plant, or animal); the simplest form of life, more than 200 viruses are known to produce human disease.

W

Water for injection (WFI) — very pure water used for medical purposes.

ACRONYMS AND ABBREVIATIONS

AIChE — American Institute of Chemical Engineers

APHIS — Animal and Plant Health Inspection Service (of the USDA)

ASM — American Society for Microbiology

ASME — American Society of Mechanical Engineers

BSL-x — biosafety level x

BSLx-Ls — biosafety level x, large scale

BOD — Biological oxygen demand

BSL — biosafety level

CBER — Center for Biologics Evaluation & Research (of FDA)

CCPS — Center for Chemical Process Safety

CFR — Code of Federal Regulations (US)

cGMP — current good manufacturing practice

cGMPs — current good manufacturing practices

CDC — Center for Disease Control

CIP — clean-in-place

COD — chemical oxygen demand

COP — clean out of place

CPI — chemical process industries

DI — deionized (water)

DNA — deoxyribonucleic acid

DOT — Department of Transportation

EMEA — European Medicines Agency

EPA — Environmental Protection Agency

FDA — Food and Drug Administration

GILSP — good industrial large scale practice

GLSP — good large scale practice

GMP — good manufacturing practice

HAZOP — hazard and operability study

HEPA — high-efficiency particulate air (filter)

HHS — Department of Health and Human Services

HPLC — high-pressure liquid chromatography

HVAC — heating, ventilating, and air conditioning

ICH — International Conference on Harmonisation of Technical Requirements for Registration of Pharmaceuticals for Human Use

ISO — International Organization for Standardization

ISPE — International Society for Pharmaceutical Engineering

L — liter(s)

MAb — monoclonal antibody

MOC — management of change

MSDS — material safety data sheet(s)

NEC — National Electric Code

NCCLS — National Committee for Clinical Laboratory Standards (NCCLS) which is now CLSI,

NIH — National Institutes of Health

OECD — Organization for Economic Cooperation and Development

OSHA — Occupational Safety and Health Administration

P&ID - piping (or process) and instrumentation diagram

PHA — process hazard analysis

PPE — personal protective equipment

PSM — process safety management

PSSR — pre-startup safety review

QA — quality assurance

QC — quality control

RDNA — recombinant DNA

RMP — risk management program

RO — reverse osmosis

SAL — sterility assurance level

SCBA — self-contained breathing apparatus

SIP — sterilization in place

SME — subject matter expert

SS — steam sanitation

TOC — total organic carbon

TSCA — Toxic Substances Control Act

UF — ultrafiltration

UK — United Kingdom

ULPA — ultra low particulate air

USDA — United States Department of Agriculture

USP — United States Pharmacopoeia

WFI — Water for Injection

WHO — World Health Organization

INDEX

A

ABSA, xix, 3, 151, 158
AIChE, xix, xxi, 152, 155, 215
American Biological Safety Association, xix, 3, 155, 156
American Institute of Chemical Engineers, iii, v, xix, xxi, 148, 151, 152, 153, 154, 155, 213
American Society for Mechanical Engineers, 3
American Society for Microbiology, 3, 155, 213
ASM, 3, 153, 155, 156, 159, 213
ASME, 3, 117, 213
Aspergillus niger, 37

B

Bacillus subtilis, 37
Behavior-based Safety, 76
Biohazard, 8, 68, 71, 130, 165, 170, ccii
Biological Hazard-based Risks, 1
Biological Safety Officer, 50, 194
Biopharmaceutical Industry, 2
Bioprocess Safety Hazards, 51, 52, 53, 54
Bioreactor, 17, 27, 29, 31, 32, 37, 39, 40, 100, 106, 118, 119, 129, 130, 132, 206
Biorisk, 54, 55, 56, 58, 59, 63, 66, 89
Biosafety Level 1, 38, 120, 121, 160, 161
Biosafety Level 2, 38, 120, 160, 163, 165
Biosafety Level 3, 38, 105, 160, 168, 169
Biosensors, 20
BL1-LS, 38, 39, 85, 100, 108, 118, 131, 161, cci
BL2-LS, 38, 39, 100, 105, 118, 138, 160, 163, cci
BL3-LS, 38, 39, 100, 108, 118, 120, 138, 139, 160, 168, 169, cci

Brevibacteruim lactofermentum, 37
BSO, 50, 160, 166, 171

C

CCPS, v, xix, xx, xxi, 48, 50, 53, 58, 59, 60, 91, 213
CDC, 3, 30, 36, 90, 139, 152, 154, 158, 159, 213
Cell Growth, 27, 29, 31, 32, 34, 35, 37
CEN, 54, 55, 59, 60, 63, 64, 152
Center for Chemical Process Safety, xix, xxi, 48, 141
Center for Disease Control, 3, 30, 213
cGMP, 215
Closed-system Design, 97, 117, 121, 120
Compliance Audit, 60
Contractors, 57, 60
Corynebacterium, 19, 37

D

Definition of a biosafety level, 83
Definition of a closed process or system, 117
Definition of bioprocess, 1
Definition of hazardous materials, 2
Definition of In vitro, 23
Definition of In vivo, 23
Definition of incident, 70
Definition of large scale, 36
Definition of small scale, 36

E

E. coli, 19, 20, 25, 31, 33, 80, 84, 133, 160
Ebola, 37, 139
Edible Vaccines, 20
Employee Participation, 60
Environmental Protection Agency, 3, 63, 151, 152, 214
EPA, 3, 59, 60, 63, 158, 212, 214
European Committee for Standardization, 54

F

FDA, 3, 63, 107, 111, 142, 158, 194, 213, 214
Food and Drug Administration, 3, 63, 142, 158, 214

G

Gene Therapy, 1, 20, 87, 141
GLSP, 38, 39, 85, 86, 108, 118, 120, 129, 130, 131, 160, 161, 176, 178, 181, cci, 214
Good Large Scale Practices, 38, 160
Good Manufacturing Practice, 3, 65, 107

H

HEPA, 42, 97, 99, 101, 103, 104, 105, 118, 120, 132, 138, 167, 172, 173, 198, 199, 214
HVAC, 91, 94, 95, 97, 99, 100, 124, 138, 139, 156, 207, 214

I

IBC, 50, 149, 158
ICH, 3, 116, 124, 214
In vitro, 208
In vivo, 208
Incident Investigation, 60
Infectious Agents, 208
Institutional Biosafety Committee, 50, 51, 160
Instructional Systems Design, 67
International Biosafety Levels, 12, 13
International Conference on Harmonization of Technical Requirements for Registration of Pharmaceuticals for Human Use, 3
International Organization for Standardization, 55, 153, 214
International Society for Pharmaceutical Engineering, 3, 214
ISO, 55, 63, 64, 97, 98, 104, 111, 153, 187, 214
ISPE, 3, 117, 214

J

Job and Task Analysis, 67

L

Learn from Experience, 60

M

MAb, 36, 211, 216
Maintenance, 4, 7, 42, 57, 62, 65, 155, 189
Manage Risk Elements, 60
Management of change, 66, 210
Management System, xxi, 2, 3, 7, 8, 11, 12, 14, 15, 47, 48, 50, 51, 54, 55, 58, 59, 61, 62, 63, 64, 65, 66, 67, 71, 76, 91, 187, 208, 209, 210
Material Safety Data Sheets, 43
Mechanical integrity, 211
Media, 6, 27, 28, 34, 35, 37, 41, 43, 80, 87, 100, 108, 119, 127
MOC, 208, 214
Monoclonal Antibody, 19, 23, 87, 209
MSDS, 43, 51, 214
Mycobacterium tuberculosis, 36, 139

N

National Institutes of Health, 3, 8, 139, 156, 215
NIH, 30, 36, 38, 50, 85, 86, 90, 118, 120, 139, 152, 154, 158, 159, cci, 215

O

Occupational Safety and Health Administration, 3, 63, 151, 152, 215
OECD, 3, 38, 156, 215
Operating Procedures, 35, 60
Operational Discipline, 76
Organisation for Economic Co-operation and Development, 3, 156
OSHA, 59, 60, 63, 64, 75, 141, 153, 154, 158, 210, 215

P

Penicillium chrysogenum, 37
Penicillium notatum, 18
Personnel Safety, 2, 48, 63, 75, 76
Pharmacogenomics, 20, 143, 209
Preliminary Design Review, 5
Process Hazard Analysis, 60
Process Safety Information, 60
Process Safety Management, xxi, 1, 2, 3, 7, 48, 58, 64, 192, 215
Product Safety, 43

Q

Quality Assurance, 7, 107, 140, 193, 215

R

Recombinant DNA, 1, 20, 25, 27, 36, 50, 86, 87, 120, 142, 153, 157, 159, 178, 179, cciv,
    209, 217
Risk Group, 30, 82, 139
Risk Management Program, 3, 14, 189, 215
Risk-based Process Safety, 59

S

Saccharomyces cerevisiae, 36, 37, 160
Safe Work Practices, 60
Scale-up, 23, 24, 27, 28, 36, 38, 43, 152
Single-use Bioprocessing Equipment, 44
Staphylococcus, 18, 84
Stem Cells, 1, 20, 141
Streptomyces, 19

T

Training, 12, 56, 60, 63, 64, 67, 165, 170
Transchromosomic, 23
Transgenic, 23, 24
Typical Hazards, 4, 5

U

Understand Hazards and Risks, 60

United States Department of Agriculture, 3, 63, 216
USDA, 3, 19, 63, 85, 145, 158, 213, 216

V

Validation, 28, 57, 97, 126, 129, 139, 166, 171
Verification, 57

W

Waste Handling, 43
Wastewater, 4, 6, 42
WHO, 3, 30, 83, 90, 139, 154, 158, 159, 216
World Health Organization, 3, 30, 151, 154, 157, 216

X

Xanthomonas, 19

Printed in the United States
By Bookmasters